Hot Isostatic Pressing
HIP'22

13th International Conference on Hot Isostatic Pressing
September 11-14, 2022, Columbus, OH, USA

Editors
Brian Welk
Victor Samarov
Cliff Orcutt
David Gandy
Hamish Fraser

Peer review statement

All papers published in this volume of "Materials Research Proceedings" have been peer reviewed. The process of peer review was initiated and overseen by the above proceedings editors. All reviews were conducted by expert referees in accordance to Materials Research Forum LLC high standards.

Published under License by **Materials Research Forum LLC**
Millersville, PA 17551, USA

Published as part of the proceedings series
Materials Research Proceedings
Volume 38 (2023)

ISSN 2474-3941 (Print)
ISSN 2474-395X (Online)

ISBN 978-1-64490-282-0 (Print)
ISBN 978-1-64490-283-7 (eBook)

This book contains information obtained from authentic and highly regarded sources. Reasonable efforts have been made to publish reliable data and information, but the author and publisher cannot assume responsibility for the validity of all materials or the consequences of their use. The authors and publishers have attempted to trace the copyright holders of all material reproduced in this publication and apologize to copyright holders if permission to publish in this form has not been obtained. If any copyright material has not been acknowledged please write and let us know so we may rectify in any future reprint.

Distributed worldwide by

Materials Research Forum LLC
105 Springdale Lane
Millersville, PA 17551
USA
https://www.mrforum.com

Manufactured in the United State of America
10 9 8 7 6 5 4 3 2 1

Table of Contents

Letter from the IHC Chairman
Committees

PM-HIP for Nuclear: Outlook, Technology and Applications
David W. Gandy 1

Near-Net-Shape HIP Manufacturing for sCO2 Turbomachinery Cost Reduction
Shenyan Huang, Victor Samarov, Dmitry Seliverstov, Jason Mortzheim,
Evgeny Khomyakov 11

Manufacturing of Compact Heat Exchangers by Hot Isostatic Pressing
Emmanuel Rigal, Isabelle Moro-Le Gall, Matthieu Maunay, Sébastien Chomette,
Lionel Cachon, Sébastien Vincent 17

Dissolvable HIP Space-Holders Enabling more Cost Effective and Sustainable Manufacture of Hydrogen Electrolyzers
Iain Berment-Parr, Owen Larkin, Bea Howarth, Kieran Bullivant 22

The Pathology of PM HIP Duplex Stainless Steels
Tomas Berglund, Björn-Olof Bengtsson, Jan-Olof Nilsson 29

Application of HIP-NNS to Large Complex Products Using Super Duplex Stainless Steel Powder
Toyohito Shiokawa, Mitsuo Okuwaki, Hiroshi Urakawa 35

Powder Metallurgy HIP for Naval Nuclear Applications – Trends in Process and Property Development
Colin D. Ridgeway, Terrance Nolan 41

Rapid L-PBF Printing of IN718 Coupled with HIP-Quench: A Faster Approach to Combine Manufacturing and Heat Treatment in a Nickel-Based Alloy
Emilio Bassini, Giulio Marchese, Davide Grattarola, Pietro A. Martelli, Sara Biamino,
Daniele Ugues 48

Use of High-Pressure Heat Treatment (HPHTTM) for L-PBF F357
Chad Beamer, Andrew Wessman, Donald Godfrey.. 55

Use of HIP Process in Post-Processing of Components Manufactured by SLM Technology from Magnetically Soft FeSi6.5 Powder
Dariusz Kołacz, Adrian Radoń, Karol Krukowski, Joanna Kulasa,
Aleksandra Kolano-Burian 61

Effect of Hot Isostatic Pressing on Microstructure and Properties of GH4169 Superalloy Manufactured by SLM
Shanting Niu, Hongpeng Xin, Lida Che, Haofeng Li, Xiangyang Li 67

Manufacturing of Net-Shape and Wear-Resistant Composite Components via the Combination of Additive Manufacturing and Hot Isostatic Pressing
Markus Mirz, Marie Franke-Jurisch, Anke Kaletsch, Simone Herzog, Yuanbin Deng,
Johannes Trapp, Alexander Kirchner, Thomas Weissgärber, Christoph Broeckmann 78

Development and Manufacture of Innovative Toughened Fine-Grained Recrystallized Tungsten Alloy
Koichi Niikura, Shunsuke Makimura, Hiroaki Kurshita, Hun-Chea Jung,
Yasuhiro Matsumoto, Masashi Inotsume, Masahiro Onoi 85

Comparison of HIP Composite and HIP Solid Material with Melting Metallurgically Produced Solid Material
Alexander Ernst, Beat Hofer, Adem Altay, Michael Hamentgen 91

Fabrication of Large Three-Dimensional Flow Path Structure Using SS Flexible Tube
Ryunosuke Kitamori, Mitsuo Okuwaki, Shigeki Tsuruoka, Takuya Nagahama 101

Powder Metallurgy HIP and Extrusion Study of FeCrAl Alloy for Accident Tolerant Fuel Cladding
Shenyan Huang, Evan Dolley, Steve Buresh, Ian Spinelli, Jason Leszczewicz,
Marija Drobnjak, Mike Knussman, Raul B. Rebak 107

Preparation of High-Quality Mo-Nb (Ti/Ni-Ti) Sputtering Target by Hot Isostatic Pressing
Zhanfang Wu, Lida Che, Jing He, Haofeng Li, Pengjie Zhang, Xiangyang Li 113

Simulation-Based Manufacturing of Near-Net-Shape Components and Prediction of the Microstructural Evolution during Hot Isostatic Pressing
Yuanbin Deng, Anke Kaletsch, Christoph Broeckmann 120

Modelling of Powder Filling of HIP Canisters
Simon Chung, Abheek Basu, Kieren Irvine, Peter Wypych, David Hastie, Andrew Grima,
Sam Moricca 131

Computational Modeling of PM-HIP Capsule Filling and Consolidation by DEM-FEA Coupling
S. Sobhani, M. Albert, D. Gandy, A. Tabei, A. Fan 141

Identification of Porous Materials Rheological Coefficient Using Experimental Determination of the Radial and Longitudinal Strain Rate Ratio
Gerard Raisson, Vassily Goloveshkin, Victor Samarov 150

A New Constitutive Modeling for Hot Isostatic Pressing of Powders
Gholamreza Aryanpour 160

HIP Modeling and Design of Large Complex Shape Parts Close to the Size of the HIP Furnace Accounting Capsules Manufacturing Technology and their Movement Inside it During the Cycle
Dmitry Seliverstov, Evgeny Khomyakov, Alex Bissikalov, Victor Samarov 166

Effect of Experimental Determination Process on Shear Stress Coefficient of Green Equation Describing HIP
Gerard Raisson, Vassily Goloveshkin, Evgeny Khomyakov, Victor Samarov 172

Theoretical Evaluation of Capsule Material Strain Hardening on the Deformation of Long Cylindrical Blanks During HIP Process
Andrey Bochkov, Yury Kozyrev, Anton Ponomarev, Gerard Raisson 177

Keywords

Letter from the IHC Chairman

Dear Colleagues,

I would personally like to thank each and every one of you for making HIP2022 a great success. When we started this journey in 2017, down in Sydney, we could have never imagined the world changing as it has, and the impact of an unforeseen pandemic. With the conference shifting multiple times to later and later dates and the unknowns of travel we were really in uncharted territory. In the end we had a great turnout with over 160 registered attendees, including the end users of the HIP products, with an excellent quantity of 55 accepted papers that required parallel sessions. There were 18 exhibitors that presented their offerings and achievements. These numbers were even attained without our usual delegations from China and other locked down countries. The event location turned out perfect, both in size and offerings, and the Ohio weather finally co-operated for once. Many attendees told me they had a wonderful time, and felt it was one of the best conferences they had ever participated in. The talks were great, the people were all so nice, and the food was good, especially the conference evening dinner. We presented multiple awards to the best part competition, the best papers, as well as two lifetime achievement awards, one to Mr. Michael Conaway, and another to Ms. Hongxia Chen. Both of these individuals have dedicated their lives to advancing hot isostatic pressing and have been members of the International HIP committee for decades.

Beginning in 2002 the IHC has regularly conducted during the HIP conferences the "Best Part Made by HIP" Competition that has always attracted a lot of interest, and HIP 2022 was not an exception. Four cutting edge developments were presented by the teams led by EPRI, UCLA, GE Global Research, and MTC Powder Solutions. The Award for the "Dual String Interlocking Shear Rams" for the off-shore applications combining complex shape and diffusion bonding of dissimilar materials was given to the MTC Powder Solutions team led by Tomas Berglund. Two lively Panel Discussions on the most important topics such as "HIP of very large parts and a road to ATLAS technology" and "HIP Modeling" were also added to the Conference Program.

The IHC also voted to hold the next conference in Aachen, Germany in 2025 with Dr. Christoph Broeckmann agreeing to oversee the event. The website is up for the next conference at www.HIP2025.com and I really look forward to seeing everyone again, as well as many new faces.

I want to thank the conference committee members of David Gandy, Dr. Hamish Fraser, Dr. Victor Samarov, and Mike Conaway for their countless hours of work planning, inviting, and qualifying the presentations. Finally, I can't say enough about the job Brian Welk did single handedly designing and updating the website, tracking the papers, planning the restaurant, handling logistics, and all the tasks needed to make this conference successful. Thank you Brian!

Sincerely,

Cliff Orcutt
Chairman
The International HIP Committee.

Committees

Cliff Orcutt
IHC and HIP2022 Conference Chairman
American Isostatic Presses

Dr. Hamish Fraser
Program Committee
The Ohio State University

Dr. Victor Samarov
Program Committee
Synertech PM

David Gandy
Program Committee
EPRI

Mike Conaway
Organizing Committee
Isostatic Forging International Group

Brian Welk
Organizing Committee
The Ohio State University

Panel Discussion on ATLAS HIP

Life Time Presentation and Award Ceremony

Best Part Contest Presentation

Best Part Award Ceremony

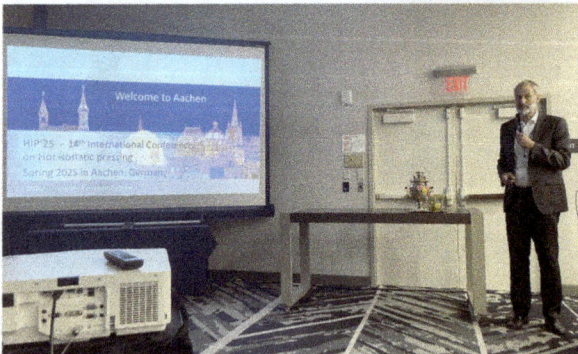

Presentation of HIP2025

HIP'22 Conference Dinner

HIP'22 Conference Dinner

HIP'22 Conference Dinner

HIP'22 Conference Dinner

PM-HIP for Nuclear: Outlook, Technology and Applications

David W. Gandy[1, a]

[1]Electric Power Research Institute, Charlotte, NC, USA

[a]davgandy@epri.com

Keywords: Hot Isostatic Pressing, Power Generation, Nuclear Energy, Modeling, Alloys, Powder Production, Large HIP

Abstract. Significant reductions in CO_2 greenhouse gas emissions must be realized to meet the US goal of a 50% overall decrease by 2030. To further meet the net-zero emission goal by 2050, substantial reductions across three primary sectors (electricity, transportation, and industrial/buildings) must also be realized. Within the electricity sector, these significant reductions can only be accomplished through the replacement of much of the existing power generation infrastructure with renewables, hydrogen, natural gas, storage and new nuclear. It is anticipated that by 2050, the US will have to replace nearly 800GW of fossil and nuclear power generation assets (Note: 1GW = ~750,000 homes or 2 coal-fired power plants). This paper highlights several planned nuclear units (40 units) that are slated for production by the 2030 timeframe. If the PM-HIP community wants to be a part of this transition and support new nuclear, it too must begin work immediately to both qualify new materials/components and further develop its infrastructure for new component manufacturing and fabrication. This paper provides an overview of the current materials that are accepted within the ASME Boiler and Pressure Vessel Code and highlights recent changes which will allow PM-HIP materials/components to be more easily integrated and accepted into the Code. Additionally, this paper identifies many of the key needs for PM-HIP to be considered part of the new build equation including two enabling technologies: PM-HIP modeling & design and large PM-HIP capabilities, along with three additional supporting needs: powder production, scaling of components, and engagement of the end-user community.

Pathway to Net-Zero

From 2005-2018, the US reduced overall CO_2 greenhouse gas emissions by ~12 percent. From 2018-2030, the US expects to reduce emissions by another 38 percent which would bring the overall reductions to 50 percent over a 25-year period. To date, much of this reduction can be attributed to the electricity industry where many coal burning fossil units have been displaced with cleaner and more efficient gas units or renewables (wind, solar, etc.). To achieve the 50% reductions by 2030 however, other industry sectors must become more deeply involved along with the electricity sector. Specifically, the transportation and industrial/buildings sectors must also reduce their emissions substantially. Hence, the big drive by industry to electrify the automotive industry and to work with industrial/building owners to significantly improve energy efficiencies. To approach net-zero applications, even further reductions will be required along all 3 sectors. Figure 1 provides a view of some of the modeling efforts by EPRI to more clearly show the reductions that will be required [1].

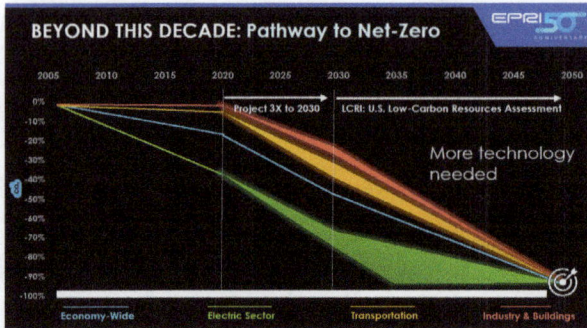

Figure 1. Pathway to Net-Zero [1].

The transformation of the electricity sector will be immense over the next four decades and will include renewables, hydrogen, natural gas, storage, and of course—nuclear (Fig. 2). It is anticipated that by 2050, the US will have to replace almost all its 800GW of fossil and nuclear generation assets. That is a huge amount of electrical capacity that will require replacement and certainly nuclear will play a large part in this replacement. Consistently over the past several decades, nuclear has made up roughly 20% (100GW) of the domestic production and is anticipated to remain around those levels. Furthermore, the 800GW does not include "new nuclear" applications that will also be added to the mix. As one can imagine, the supply chain will have to significantly ramp upwards to accommodate the demand that is expected over the next several decades. More on this topic can be found in References 1 and 2.

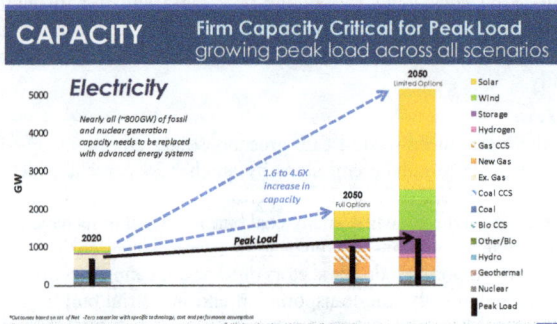

Figure 2. Project electricity capacities within North America through 2050 [1].

Advanced Nuclear Deployment Plans

The Nuclear Energy Institute (NEI) recently highlighted several planned or under construction projects in the US and Europe (note, this doesn't include China and Russia where significant investments are also being made) [3]. Table 1 provides a snapshot of ~40 reactors (~7500MW total) that are planned to be operational by 2030-32. These units include grid-scale reactors and represent only the tip of the iceberg in terms of new builds over the next decade. A similar list can also be found for micro-reactors (reactors below ~10MW) [3]. Even at this level, it is anticipated that the supply chain will be strained to meet the demand as much of the manufacturing/fabrication capabilities in the USA have now moved overseas, NEI suggests that as many as 300 reactors generating 90GW of "new capacity" might represent the low end of commissioning over the next three decades [4].

Table 1. Advanced reactor deployment plans [3].

Developer	Utility / User	Location	Size	Target Online
NuScale	UAMPS	Idaho, USA	6 @ 77MW	2029
	KGHM Polska Miedz	Poland	6 @ 77MW	2029
	Nuclearelectrica	Romania	6 @ 77MW	2028
GEH BWR X-300	OPG	ON, Canada	300 MW	2028
	TVA	TN, USA	300 MW	2032
	Synthos & Orlen	Poland	300 MW (>10 plants)	Early 2030s
	SaskPower	Sask., Canada	~300 MW (4 plants)	2032 to 2042
Holtec SMR-160	TBD	NJ, USA	160 MW	2030
X-energy Xe-100	Grant County PUD	WA, USA	4 @ 80MW	2027
TerraPower	Pacific Corp.	Wyoming	345 - 500MW	2028
ARC	NB Power	NB, Canada	100 MW	2030
Moltex	NB Power	NB, Canada	300 MW	2032
TBD	Purdue/Duke Energy	Indiana, USA	TBD	TBD

So as one can see, "WE'RE GOING TO BUILD A LOT OF POWER PLANTS AND COMPONENTS OVER THE NEXT SEVERAL DECADES!"
Even if these projections are under by 50%, there is still ample space for the PM-HIP community to be deeply involved. Later in this paper, some "key needs" and common areas that the HIP community can begin working on collaboratively will be presented.

Approved Alloys Added to 2021 Edition of ASME Code
The ASME Boiler and Pressure Vessel Code has been developed over many decades by industry to provide standard rules for the construction of steam boilers and other pressure retaining components. Section I of the Code provides rules governing Power Boiler applications, Section II provides rules for Materials, and Section III provides rules for Construction of Nuclear Facility Components, while Section VIII provides rules for Pressure Vessels. Until recently, PM-HIP was only acknowledged within the ASME Code through a handful of Code Cases. In 2021, Section II-Materials recognized ~30 HIP materials for the first time by incorporating several ASTM specifications as SA/SB standards:

- SA988/SA988M -- Specification for Hot Isostatically-Pressed Stainless Steel Flanges, Fittings, Valves, and Parts for High Temperature Service
- SA989/SA989M -- Specification For Hot Isostatically-pressed Alloy Steel Flanges, Fittings, Valves, And Parts For High Temperature Service
- SB834/SB834M -- Specification for Pressure Consolidated Powder Metallurgy Iron-Nickel Chromium-Molybdenum (UNS N08367), Nickel-Chromium- Molybdenum Columbium (Nb) (UNS N06625), Nickel- Chromium-Iron Alloys (UNS N06600 and N06690), and Nickel-Chromium-Iron-Columbium Molybdenum (UNS N07718) Alloy Pipe Flanges, Fittings, Valves, and Parts

This recognition is very significant in that several austenitic stainless steels, ferritic steels, and nickel-based alloys are now available for use in pressure retaining applications.
Additionally, and equally important, new guidance was provided to industry for qualification and acceptance of PM-HIP. This permits PM-HIP to be used for the manufacture of components in a

3

similar manner to that applied for forged, cast, or other wrought product forms as long as one can qualify the material. The guidance is provided under the following:

- BPV-II, Part D, Mandatory Appendix 5 -- Guidelines on the Approval of New Materials Under the ASME Boiler and Pressure Vessel Code

For materials accepted under Section II-Materials to be used in pressure retaining applications, one of the three Book Sections (Section I, III, VIII) must also recognize the material. For nuclear applications, Section III has recently incorporated/recognized 316L stainless steel (UNS S31603) under a Mandatory Appendix (Record No. 21-2331) that permits its use for component manufacture. Prior to this, 316L SS was only recognized under Code Case N-834. The incorporation of 316L SS now provides a blueprint for recognition of additional alloys under Section III. Priority alloys which may be incorporated over the next several years include: SA508 low alloy steel and several nickel-based alloys: 600, 617, 625, 690 and 800H.

Approved Alloys & New Alloys Needed
Many of the alloys found within nuclear applications to date have been manufactured using product forms such as forgings, castings, extrusions, etc. Applications have been for the most part at reasonably lower temperatures (<400C) to date. As industry moves toward higher temperature (550-750C) Advanced Reactor applications, additional alloys qualifications will be required. To date, only six alloys have been recognized for high temperature nuclear applications:

- 2-1/4Cr-1Mo
- A508 Grade 3 Class 1 and SA533 Type B Class1
- 9Cr-1M-V (Grade 91)
- 304/304H and 316/316H
- Alloy 800H
- Alloy 617

Many additional alloys are currently being considered for nuclear applications or actual qualification of various product forms of these alloys is underway. EPRI has developed an Advanced Reactor Materials Development Roadmap that highlights many of these alloys. (5). Several of these are highlighted below:

- Stainless Steels
- 316LN, 316H, 316FR, 15-15Ti, D9, Alumina forming SS
- Ferritic or Ferritic-Martensitic Alloys
- 508 Grade 3, Classes 1 and 2
- F/M-9Cr and 12Cr
- Nickel Alloys
- 625, 690, 617, 800H
- Cladding applications
- Mo, W, Hastelloy N

As one might anticipate, many of these alloys can be readily produced by the PM-HIP process based on current industry experience. It's simply a matter of qualifying the alloys for Section III, Division 5 applications. The PM-HIP industry is encouraged to work with EPRI and various OEMs to bring these alloys forward for higher temperature service.

What's Required for PM-HIP to Be Part of the Plans for Advanced Nuclear Deployment – Key Needs

As noted earlier, the nuclear industry plans to deploy ~40 SMRs and ARs by the 2030 timeframe. To accomplish this, many have elected to use conventional product forms for first-of-a-kind (FOAK) applications. However, as new manufacturing methods and alloys are qualified, OEMs will look toward more advanced manufacturing methods such as PM-HIP and Additive Manufacturing to produce components. To be part of the overall plans for advanced nuclear deployment, the PM-HIP community must begin qualification of components/materials now as it can often take 3-5 years to gain acceptance within the Code. Furthermore, there are several other key needs that must be addressed over the next few years if PM-HIP is to become mainstream for nuclear applications:

- PM-HIP Modeling & Design
- Powder Production (quality & quantity)
- Large HIP
- Scaling of Components
- Engagement of End-User Community

Each of these needs will be discussed further below. However, before beginning that discussion, let's look a bit more deeply at what the cost drivers are for PM-HIP applications. Figure 3 provides a good overview of both the cost drivers and the commercial availability for each step in the overall PM-HIP process. As can be seen from the figure, modeling/design of the capsule and powder costs are the two most significant influencers from a cost perspective, while model/design and capsule filling are the two steps that are currently very limited in terms of commercial availability. Each of these elements play a large role in acceptance of PM-HIP technology for the production of large parts and require further focus by industry.

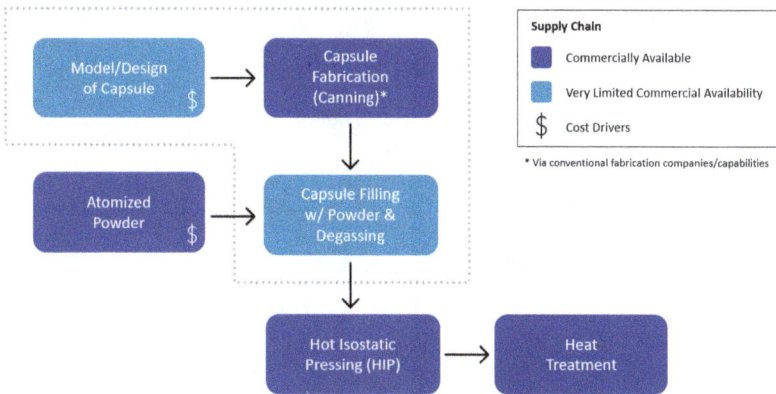

Figure 3. Cost drivers and commercial availability for each step in the overall PM-HIP process.

PM-HIP Modeling & Design. Modeling & capsule design are seen as a key enabler for PM-HIP technology expansion and deployment. Today only a handful of modelers/designers currently exist in this market around the world. This is very limiting if the industry plans to expand into large component production. If industry expects PM-HIP technology to expand further and become "mainstream" to compete (and/or supplement) forging and casting technologies, industry must develop improved modeling capabilities and performance. Other forming industries use advanced tools like DEFORM®, MAGMA®, and PROCAST® to name a few, that enable engineers to predict casting and forging product sizes and movement. A similar capability for modeling/design is sorely needed for the PM-HIP industry if modelers/designers of equipment are expected to consider PM-HIP for various pressure retaining and large structural applications. Constitutive models have been developed for <20 alloys to date. (6-15) Furthermore, constitutive properties are often held by the developer and used as a competitive advantage. Industry should look to come together to share models/properties and to develop new models/properties for alloys that could be used soon. Development of software models/properties are the #1priority for the PM-HIP industry to be a part the advanced nuclear equation.

Powder Production. Over the past decade, many powder manufacturers have moved to support the additive manufacturing community. Unfortunately, this has led to a flat or possibly lower powder production rate for PM-HIP applications, just as the nuclear industry is looking to dramatically expand into new alloys/markets. Powder manufactures are encouraged to work with OEMs and developers to better understand the needs in powder production for nuclear applications and to support development of new alloy powders for this industry.

The following example demonstrates how one organization could use PM-HIP for nuclear applications. Please note, this represents only one OEM, and many others are going to be part of the nuclear market. The NuScale Power reactor is used here only as an example. Table 2 identifies four major components (lower head, upper head, steam plenum, and access ports) that could be produced from PM-HIP, while the rest of the reactor would be manufactured with conventionally forged products. As shown, the total weight for these four components would approach 100,000lbs (45,300kgs) for just one reactor. At full production, NuScale Power hopes to produce as many as 12 reactors annually. This would significantly strain the current powder production capabilities of industry today. It is anticipated that other OEMs would expect similar production capacities.

Table 2. An example of the projected annual powder requirements for one OEM.

Weights (full-scale) for <u>one</u> reactor:
- Lower Head – 19,000lbs
- Upper Head – 21,000lbs
- Steam Plenum – 39,000lbs
- Access Ports – 5100lbs x 4

➡ ★ ~100,000lbs/unit X 12/year

Large HIP. Two large scale HIP efforts, ATLAS and TITAN, are being considered by industry to produce PM-HIP components that are >3.0m in diameter. Specifically, ATLAS-4.05m and TITAN-~4.6m are both under consideration. ATLAS (Fig. 4) appears to be slightly ahead in terms of design/deployment now, but both appear to be serious considerations. For HIP to be part of the nuclear equation, one or more of these systems must be designed, fabricated, and commissioned. Both will more than likely exceed $200M (if building costs are included) and will more than likely include some level of government funding.

Hot Isostatic Pressing - HIP'22   Materials Research Forum LLC
Materials Research Proceedings 38 (2023) 1-10   https://doi.org/10.21741/9781644902837-1

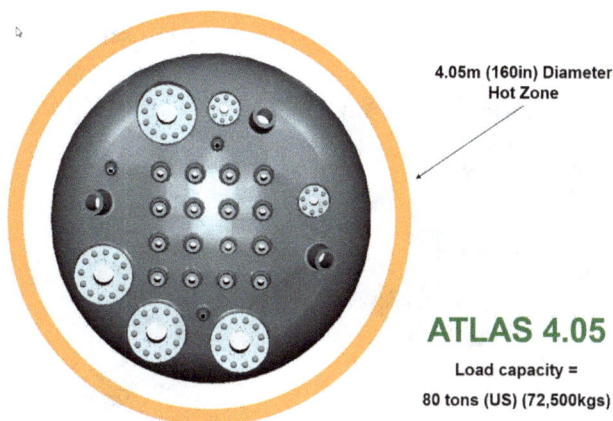

Figure 4. A schematic showing a reactor head within ATLAS 4.05.

Current plans suggest that ATLAS will be funded by a Consortium of industry investors and will be more than likely led by the BWXT, UNNPP, Stack Metallurgical, and EPRI, while the UK effort will be led by Rolls-Royce and other partners. The HIP 2022 conference will hopefully provide some greater clarity around the scope and plans of each "Large HIP" application.

Scaling of Components. Another key and often overlooked need for HIP to be realized for nuclear applications is scaling of the technology for larger components. Many of the components anticipated currently exceed the capacity of today's HIP units. As industry moves to scale beyond ~1.25m (50 inches) in diameter, many of these components will need to be scaled to assure both properties and dimensional conformance. One example that is currently being pursued by the US Department of Energy project on SMRs Manufacture and Fabrication (16) is the production of both an upper and lower head using PM-HIP. Initial research was performed on a 44% scale upper head (which would just fit inside a 65-inch HIP) (see Fig. 5). Next, it is being scaled to a 2/3-scale by producing the head in half sections and then welding it together. Eventually, when ATLAS or TITAN become realities, the heads could be produced in full section. So as one can see, the process requires some progression in scaling. It also requires development of properties at such a large scale.

Materials Research Forum LLC
https://doi.org/10.21741/9781644902837-1

Figure 5. Scaling of large components such as a reactor head may take several iterations to reach acceptable dimensional and property requirements. A 44% upper head and one-half of a 67% upper head were manufactured under a large US DOE project [16].

Engagement of End-User Community. Lastly, production of large, nuclear components requires engagement from the end-user community as well as power producers, designers, and HIP suppliers. The end-user community is made up of major OEMs and fabricators as well as the utilities that purchase the nuclear units. Engagement is paramount for the technology to be even considered for nuclear applications. EPRI recently conducted a Supply Chain Workshop for Advanced Energy Systems which brought together both end-users and manufacturers, fabricators, etc. (2). A follow up workshop is planned for Q1-2023 to continue bringing interested participants together to address industry needs. Additionally, EPRI is also working with numerous industrial partners and an EPRI Board-funded Initiative to support Advanced Manufacturing, Methods and Materials (AM3) to accelerate qualification and deployment of high temperature materials and manufacturing processes (Fig. 6).

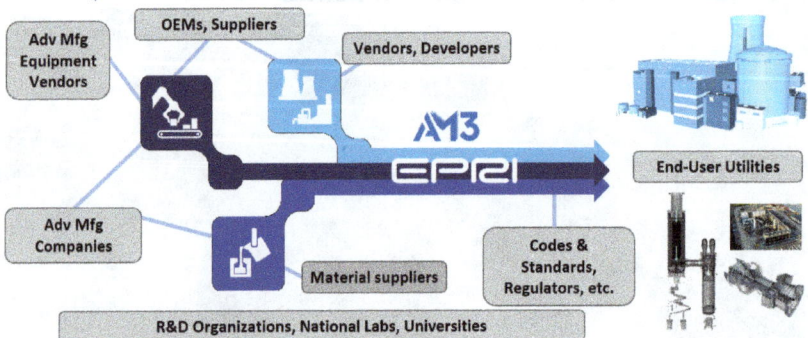

Figure 6. Engaging the entire supply chain will accelerate technology adoption.

Materials Research Forum LLC
https://doi.org/10.21741/9781644902837-1

Summary

This paper has provided discussion around various needs and activities necessary for PM-HIP to be considered part of the equation for Small Modular Reactor and Advanced Nuclear Reactor applications. Specifically, the paper introduced both a Pathway to Net-Zero (which was based on significant EPRI modeling efforts) and discussed Advanced Nuclear Deployment Plans. From these efforts, it is clear that over the next four decades (by 2050) that approximately 800GW of existing fossil and nuclear production will need to be replaced by renewables, hydrogen, natural gas, storage, and nuclear in North America alone. It is anticipated that nuclear units will account for more than ~100GW or more of this replacement. This doesn't even include new nuclear applications; this only addresses replacement.

The paper also provided a snapshot of many of the alloy qualification needs and highlighted some of the ASME Code changes that have occurred recently to allow production of PM-HIP pressure retaining components. Priority alloys which require qualification and incorporation over the next several years include: SA508 and several nickel-based alloys: 600, 617, 625, 690 and 800H.

Finally, the paper reviewed some of the key needs around what is required for PM-HIP to be part of the plans for advanced nuclear deployment. The five key needs discussed herein include:

- Modeling & Design
- Powder Production
- Large HIP Capability
- Scaling of Components
- Engagement of the End-User Community

The two key enablers include: Modeling & Design and Large HIP Capability. Today only a handful of modelers/designers currently exist within the PM-HIP community around the world. This is very limiting if the industry plans to expand into large component production. If industry expects PM-HIP technology to expand further and become "mainstream" to compete (and/or supplement) forging and casting technologies, industry must develop improved modeling capabilities and performance which allow designers and manufacturers to rapidly model/design capsules for production purposes.

Two large scale HIP efforts, ATLAS and TITAN, are being considered by industry to produce components that are >3.0m in diameter. Specifically, ATLAS-4.05m and TITAN-~4.6m are both under consideration and are highlighted herein.

References

[1] N. Wilmshurst, Economy Wide Decarbonization, Supply Chain Opportunities, EPRI Supply Chain Workshop, Dallas, TX, June 15, 2022.

[2] Supply Chain Challenges and Opportunities for Structural Components in Advanced Energy Systems: EPRI Workshop Summary. EPRI, Palo Alto, CA: 2022, EPRI Report 3002025254.

[3] H. Lane, North American Roadmap—Nuclear Energy Institute, EPRI Supply Chain Workshop, Dallas TX, June 15, 2022.

[4] K. Silverstein, The Inflation Reduction Act Will Spawn Nuclear Energy's Growth, Forbes Magazine, August 22, https://www.forbes.com/sites/kensilverstein/2022/08/22/the-inflation-reduction-act-will-spawn-the-growth-of-nuclear-energy/?sh=2b6f0c5b4158

[5] Advanced Reactor Materials Development Roadmap, EPRI Report 3002022978, October 2021,https://www.epri.com/research/products/000000003002022979

[6] Van Nguyen et al. A combined model to simulate the powder densification and shape changes during hot isostatic pressing, Computer Methods in Applied Mechanics and Engineering, 2017 Volume 315, Pages 302-315.

[7] Kohar et al. A new and efficient thermo-elasto-viscoplastic numerical implementation for implicit finite element simulations of powder metals: An application to hot isostatic pressing, International Journal of Mechanical Sciences Volume 155, 2019, Pages 222-234

[8] K. Kim and H. Yang, Densification behaviour of titanium alloy powder under hot isostatic pressing. Powder Metallurgy. Volume 44, 2001 - Issue 1, Pages 41-47

[9] D. Lasalmonie et al. Hot Isostatic Pressing of SY 625 Powder, Superalloys Conference, 1998

[10] Zhou et al, Numerical Simulation and Optimization of the hot isostatic pressure process of a part of aircraft structure, Procedia manufacturing 37 (2019), 138-145

[11] Essa, Khamis, et al. An iterative approach of hot isostatic pressing tooling design for net-shape IN718 superalloy parts. The International Journal of Advanced Manufacturing Technology 83.9-12 (2016): 1835-1845.

[12] Svoboda, Ales, L. Bjork, and H. A. Haggblad. "Determination of material parameters for simulation of hot isostatic pressing." WIT Transactions on Modelling and Simulation 12 (1970).

[13] Atkinson, H. V., and S. Davies. "Fundamental aspects of hot isostatic pressing: an overview." Metallurgical and Materials Transactions A 31.12 (2000): 2981-3000.

[14] Klar, Erhard, and Prasan K. Samal. Powder metallurgy stainless steels: processing, microstructures, and properties. ASM international, 2007.

[15] Beiss, P., and M. Dalgic. "Structure property relationships in porous sintered steels." Materials Chemistry and Physics 67.1-3 (2001): 37-42.

[16] EPRI/DOE Report 3002019335, Small Modular Reactor Vessel Manufacture and Fabrication—Phase 1 Progress (Year 2), Technical Update, April 2020.

Hot Isostatic Pressing - HIP'22
Materials Research Proceedings 38 (2023) 11-16

Materials Research Forum LLC
https://doi.org/10.21741/9781644902837-2

Near-Net-Shape HIP Manufacturing for sCO2 Turbomachinery Cost Reduction

Shenyan Huang[1,a*], Victor Samarov[2,b], Dmitry Seliverstov[2,c], Jason Mortzheim[1,d], Evgeny Khomyakov[2,e]

[1]GE Research, 1 Research Circle, Niskayuna, NY 12309, USA

[2]Synertech PM, Inc., 11711 Monarch Street, Garden Grove, CA 92841, USA

[a]huangs@ge.com, [b]Victor@synertechpm.com, [c]Dmitry@synertechpm.com, [d]jason.mortzheim@ge.com, [e]Evgeny@synertechpm.com

Keywords: Near-Net-Shape HIP, Superalloy, sCO2 Turbomachinery, Mechanical Property

Abstract. sCO2 turbomachinery that operates above 650°C requires the use of γ' strengthened Ni-based superalloys, leading to high cost and barrier of market adoption. Near-net-shape (NNS) HIP manufacturing with 282® alloy powder is being developed for sCO2 turbine components, with a significant estimated cost reduction. Tensile, creep, low cycle fatigue properties of argon gas atomized and plasma atomized powders were evaluated and compared to sand cast HAYNES® 282®. While tensile strength and fatigue life outperformed sand cast material, a 10~25% debit in the creep stress capability was observed due to the fine grain size and presence of prior particle boundaries (PPBs). Finite element model calibrated by powder rheological properties accurately predicted the nonuniform shrinkage during HIP, providing HIP tooling design to achieve the target dimension. A 20 lbs. turbine nozzle ring was successfully demonstrated within 0.01 inch dimensional tolerance at the vanes. A 1700lbs. turbine casing with complex internal struts and manifolds was also demonstrated being close to the target dimension.

Introduction

Supercritical carbon dioxide (sCO2) turbomachinery that operates above 650°C requires the use of γ' strengthened Ni-based superalloys, leading to high component cost and barrier of market adoption. Despite the 10X smaller size of a sCO2 power cycle than a steam power cycle, the LCOE of sCO2 power cycle in concentration solar-thermal power (CSP) remains to be a significant portion. HAYNES® 282® alloy with superior creep strength is a candidate for sCO2 turbine components. Supply chain is limited for large forging or casting in HAYNES® 282®. Initial attempt to sand cast a multi-piece turbine casing was not successful because of casting defects that required extensive weld repair, while extensive machining associated with massive material waste and high cost would be required if starting with a large forging. Therefore, alternative manufacturing route of powder metallurgy (PM) NNS HIP is being explored to reduce the cost of static turbomachinery components. In the present work, mechanical properties of PM HIP 282® alloy, NNS HIP feasibility of turbine nozzle ring and casing, technoeconomic impacts were evaluated.

Mechanical Properties of PM HIP 282® alloy

Argon gas atomized (GA) and plasma atomized (PA) 282® alloy powder were acquired for evaluating tensile, creep, and low cycle fatigue (LCF) properties of PM HIP 282® alloy. Despite the higher cost than GA powder, PA powder has the advantages of improved cleanliness, highly spherical morphology, low trapped gas/porosity, high yield in the powder size range appropriate for PM HIP. GA powder used was 53~150um in size with 62% relative tap density. PA powder used was 20~180um in size with 67% relative tap density. PA powder contains lower oxygen than

GA powder (compositions shown in Table 1). HIP condition of 1204°C/15ksi/4hr was performed on lab-scale HIP cans. The resulting average grain size measured by EBSD was 17um and 22.6um for GA and PA powder, respectively. A representative PA powder HIP microstructure is displayed in Figure 1d. Titanium rich MC carbonitrides and aluminum rich oxides were observed to decorate the PPBs and effectively pin grain boundary motion. The lower oxygen in PA powder is believed to introduce lower amount of PPB particles and thus leading to a slightly larger grain size. Solution treatment (1149°C/1hr/WQ) followed by two-step aging treatments (1010°C/2hr/WQ + 788°C/8hr) was applied after HIPing.

Mechanical properties of PM HIP 282® alloy were tested at relevant operating conditions and compared to wrought, sand-cast, and forged materials in literature [1-2]. As plotted in Fig. 1, PM HIP reveals significantly higher yield strength and ultimate tensile strength (UTS) than sand-cast and even higher than wrought, possibly owing to its fine grain size. PM HIP shows similar magnitude of tensile elongation with sand-cast, but 10~15% lower than wrought. Each data point represents an individual test. A smaller sample-to-sample variation in PM HIP than sand-cast material is noticed. PA powder slightly outperforms GA powder in tensile strength and ductility at elevated temperatures. Creep rupture tests were conducted at 1400°F to generate the Larson-Miller plot in Fig. 2a. A 10~15% debit in the creep stress capability is observed in PA powder HIP vs. wrought or sand-cast materials. The limited data of GA powder HIP at high creep stress indicate its inferior creep resistance than PA powder. The PPB network and finer grain size are likely to contribute to the lower creep ductility and faster creep rates in PM HIP materials. Both GA and PA powder HIP exhibits fatigue property superior to sand-cast and wrought materials (Fig. 2b). GA powder HIP shows slightly longer LCF life than PA powder at 0.7% and 0.6% strain ranges, while similar life is observed at higher strain range of 0.9%. PA powder was selected for NNS HIP part demonstration given the better creep resistance than GA powder.

Table 1. PA and GA 282® alloy powder compositions vs. nominal composition

(weight percent, wt.%)

Alloy	Ni	Cr	Co	Mo	Al	Ti	C	B	N	O
Nominal	Bal.	20	10	8.5	1.5	2.1	0.06	--	--	--
PA powder	Bal.	19.4	10.4	8.4	1.6	2.1	0.06	0.003	0.006	0.007
GA powder	Bal.	19.3	10.2	8.3	1.6	2.1	0.05	0.004	0.006	0.013

Turbine Nozzle Ring Demonstration

A 20 lbs. turbine nozzle ring was first selected to demonstrate the feasibility of NNS HIP with 282® alloy. Powder rheological properties were measured to calibrate the finite element HIP model, including specific heat, density, coefficient of thermal expansion, thermal conductivity in fully densified material up to the HIP temperature, yield stress as a function of density in fully and partially densified material. The HIP tooling design (Fig. 3a) was guided by the HIP simulation to achieve net shape at the vanes. The manufactured HIP tooling with carbon steel (Fig. 3b) had an insert (Fig. 3c) placed into the canister with features to position the nozzle ring correctly. The insert being a separate piece allowed more conventional machining operations to be used. After the tooling was assembled and welded, PA powder was filled through the stems, outgassed (Fig. 3d), and helium leak checked. The part went through HIP cycle, pre-machining to produce a sonic shape for immersion ultrasound inspection (Fig. 3e), and acid leaching in a HCl based solution to remove the carbon steel tooling and reveal the nozzle ring (Fig. 3f). After a first nozzle ring was fabricated, trailing edge defects (traces of machining tool, incomplete profile of trailing edge) were observed, which originated from vibration/bending/deflection of the needle tool during the insert slot machining. HIP model of nonuniform shrinkage and HIP tooling design were validated by the resulting dimensional tolerance of the first nozzle ring (vane within ±5mil, hub and shroud within ±10mil). Improvements were made in a second nozzle ring by enlarging the trailing edge to allow the use of a larger deflection-free tool in 5 axis milling. Chemical milling was added after acid leaching to meet the target dimension as well as removing the interdiffusion layer at the can and powder interface. The second nozzle ring successfully eliminated the trailing edge defects and tool marks (Fig. 3g, 3h). All vanes were of exceptional uniformity and repeatability. Dimensional inspection (Fig. 3i) shows a vane contour close to the target dimension with all the surfaces 15~20mil undersized except trailing edge. The problem was identified as faster chemical milling rate on the vanes than the pre-machined shroud surface. It was speculated that the powder profile and the interdiffusion layer of the vane surface could be more susceptible to chemical milling. Average surface roughness was Ra 6.2μm on pressure side and Ra 3.7μm on suction side of the vanes, suggesting that superfinish (a well-established process) would be required to achieve the target roughness of Ra 3.175μm. These encouraging results and learnings were then applied to the turbine casing NNS HIP demonstration.

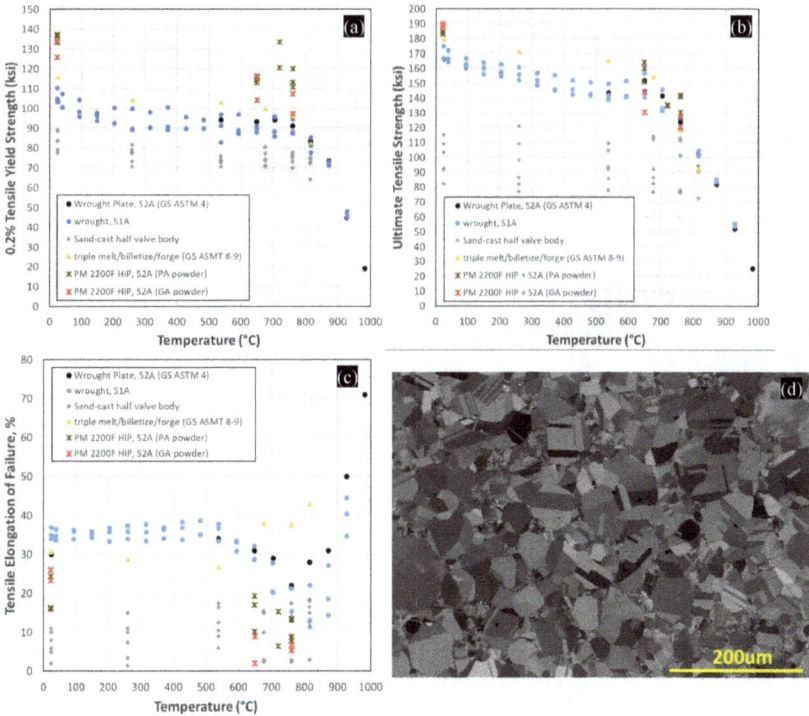

Figure 1: (a) 0.2% yield strength, (b) UTS, and (c) tensile elongation of PM HIP 282® alloys in comparison with wrought, sand-cast, and forged HAYNES® 282®. (d) Typical HIP microstructure of PA powder in a backscattered electron image.

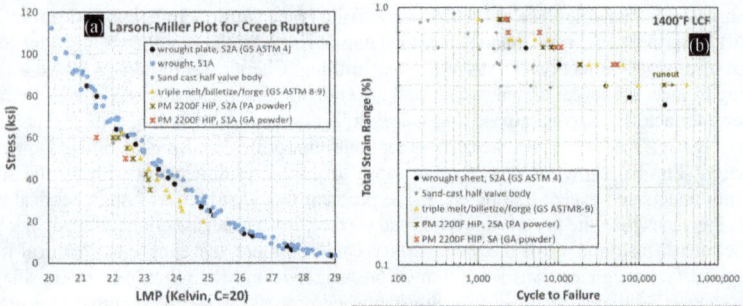

Figure 2: (a) Larson-Miller plot for creep rupture and (b) 1400°F LCF properties of PM HIP 282® alloys in comparison with wrought, sand-cast, and forged HAYNES® 282®.

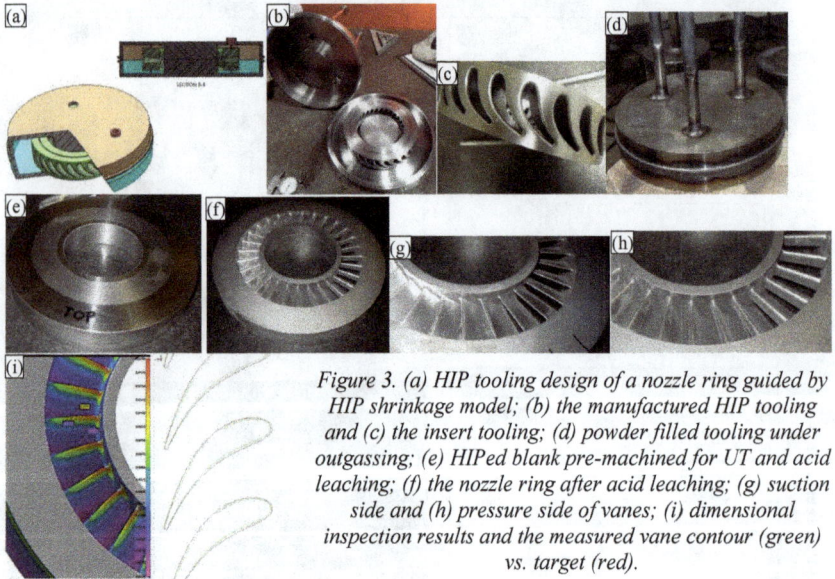

Figure 3. (a) HIP tooling design of a nozzle ring guided by HIP shrinkage model; (b) the manufactured HIP tooling and (c) the insert tooling; (d) powder filled tooling under outgassing; (e) HIPed blank pre-machined for UT and acid leaching; (f) the nozzle ring after acid leaching; (g) suction side and (h) pressure side of vanes; (i) dimensional inspection results and the measured vane contour (green) vs. target (red).

Development of the 282® alloy turbine casing via PM NNS HIP had to address several technical challenges caused by the new material, size, shape and internal geometry, including HIP tooling design and fabrication (Fig.4a-c), powder filling/outgassing, HIP (Fig. 4d), dimensional control, NDT inspection, de-canning (Fig.4f). With the powder weight in the HIP can exceeding 1700 lbs., a massive non-uniform shrinkage (exceeding 10% linear, or several inches per main dimensions) had to be predictable and considered in the HIP tooling design. The non-uniform shrinkage was influenced by the powder rheology properties, plastic stiffness of the HIP can during HIP, the 3D effects caused by the external ports, and most important, internal tooling elements (inserts) forming the manifolds and the struts. The HIP tooling manufacturing involved several techniques of sheet material fabrication and very precise 3D CNC machining that had to be matched in assembling and welding to provide a perfect fit, avoid distortions, ensure a vacuum tight assembly that will survive the 2200°F HIP with severe deformation of the welds. Despite a good flowability of the

282® PA powder, filling a volume with numerous inter-connected cavities required an intricate filling system to ensure uniform filling. Most of the filling ports also served as multiple outgassing exit channels, but still due to the molecular flow of gases inside the powder bulk, high vacuum combined with elevated temperature and time was mandatory to remove the physically adsorbed air and moisture from the surface of the powder particles to ensure full consolidation and further re-crystallization. HIP is a well-established process but has its own "under-water stones" when large PM parts, taking most of the furnace volume are processed. With the excellent uniformity of temperature at the hold stage due to high density of argon, the ramp stage of the HIP trajectory has usually high temperature gradients along the height causing earlier deformation of HIP capsules in their upper part. This was also accounted in the HIP tooling design to minimize distortions. As a result, in addition to reaching the 100% density (confirmed by the TIP tests and microstructure), the turbine casing adequately reproduced the shape from the FEM simulation (Fig. 4g) with small distortions and deviations found after scanning and dimensional analysis can be amended within the 2nd iteration of the tooling. During HIP the capsule perfectly bonds to the Ni base powder "substrate", this has allowed to produce ultrasound inspection with the HIP tooling on, when the internal volumes are still "solid", and to inspect then "through the steel" the intricate internal features of the manifolds with the two arrays of vanes/struts. The HIP tooling removal was done by acid leaching and presented several challenges, mainly due to the weight and size of the parts and the amount of the steel to be removed. Due to the HIP tooling design leaving the main inserts hollow and vented during HIP, it was possible to provide an efficient tooling removal from the vanes and manifolds without "dead" zones inside that could cause gaseous products and chemical attack of the alloy. While the outside geometry of the turbine casing was accessible for the final machining, the internal geometry was not and the goal was to provide the "net shape" of the 3D structs and manifolds. The dimensional analysis done after sectioning revealed a good correspondence and the possibility to reach 20~40mil tolerances for the 2nd iteration.

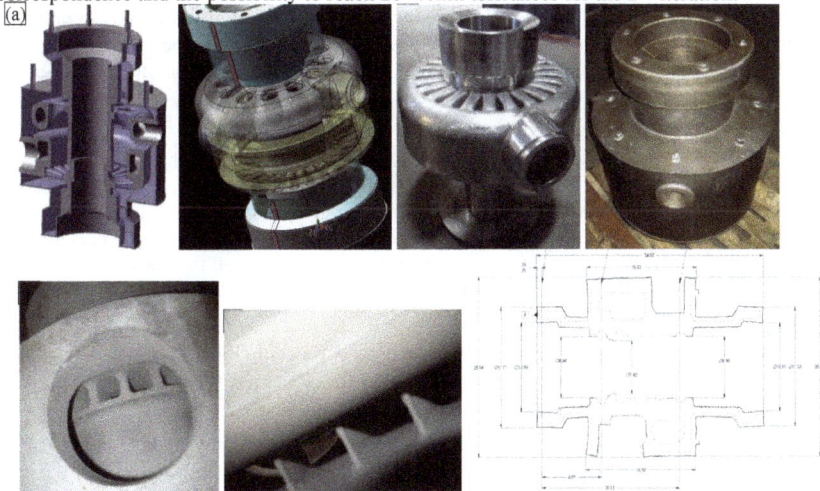

Figure 4: (a) CAD model of turbine casing HIP tooling; (b) CAD model of two internal inserts assembled with ID steel tooling; (c) assembly of internal vented tooling to form manifolds and struts; (d) HIPed part; (e) and (f) internal struts free of carbon steel tooling after acid leaching; (g) the HIP deformation map of the turbine casing.

Technoeconomic Analysis

A detailed cost model of NNS HIP based on US manufacturing was performed for the nozzle ring and compared to current manufacturing processes and the actual hardware costs for the 3 stages of nozzle rings that make up the turbine under the STEP program (DE-FE0028979). Those nozzle rings were created from forgings, with initial rough machining using conventional multiaxis mills followed by multi-axis plunge EDM operations to create the vanes. As the STEP nozzle rings were not produced from HAYNES® 282® alloy, material cost adjustment was made. The estimated cost for forged and machined complete set of 3 nozzle rings is $408,124 ($33,124 in material, $375,000 in machining). For the NNS HIP process, the costs of manufacturing 3 nozzle rings are broken into powder cost of plasma atomized powder in small lot ($3,933), Non-Recurring Engineering (NRE) ($30,000), NNS HIP process including HIP tooling, powder filling/outgassing, HIP cycle, acid leaching ($83,500), post-processing including heat treatments and final machining ($22,500). The total cost estimate for NNS HIP ($139,933) is only 34% of the conventional. 20 STEP turbines were considered to show the volume cost reductions. Costs reduce slightly as the NRE costs are amortized over a larger volume as well as a small reduction in NNS HIP costs. The NNS HIP costs are not reduced significantly due to the current HIP conditions that require a HIP unit which has a limited volume capacity and therefore can only hold 2 turbines worth of stages in a single HIP operation. The cost estimate becomes $110,183 per turbine, suggesting a 21% cost reduction on volume basis compared to a single unit. The NNS HIP process cost is dominant, well over 3x the post processing costs, the 2nd highest. NNS HIP manufacturing provides a significant LCOE reduction (> 75% $/kW reduction) over the conventional manufacturing for sCO2 turbine.

The impact of creep debit in PM HIP 282® alloy can be offset by increasing the wall thickness of the turbine casing only at the creep limiting locations, per ASME BPVC code Part III Section D. A 13% wall thickness increase is required to match the creep capability of sand-cast material, resulting in 35lbs additional powder and a marginal effect on the cost. On the other hand, pipe components made by NNS HIP are projected to have a larger cost penalty due to creep property, given the uniform wall thickness. Comparing PA and GA powder for a pipe fitting component, GA powder would need 12% wall thickness increase (30% more powder mass) than PA powder due to its creep deficiency.

Summary

NNS HIP fabrication with 282® alloy powder was demonstrated to be technically feasible and economically viable for sCO2 turbine components (nozzle ring and turbine casing). Dimensional tolerance was met for the vanes and complex internal struts and manifolds. Despite the superior tensile and fatigue properties, the creep debit in PM HIP material is intrinsic to its PPB network and fine grain structure. A cost-performance tradeoff by increasing wall thickness to compensate creep is highly dependent on the design requirements of the components.

Acknowledgement

This work was funded by US Department of Energy under award DE-EE0008996. The authors thank Haynes International for HAYNES® 282® wire and argon gas atomized powder through Praxair, AP&C (GE Additive) for plasma atomized powder, TechMet for acid leaching.

References

[1] R. Purgert, J. Shingledecker, D. Saha, M. Thangirala, G. Booras, J. Powers, C. Riley, H. Hendrix, Materials for Advanced Ultrasupercritical Steam Turbines, DE- FE0000234 Final Report, 2015. https://doi.org/10.2172/1243058

[2] B.A. Pint, H. Wang, C.S. Hawkins, K.A. Unocic, Technical Qualification of New Materials for High Efficiency Coal-Fired Boilers and Other Advanced FE Concepts: Haynes® 282® ASME Boiler and Pressure Vessel Code Case, ORNL/TM-2-2-/1548 Report, 2020.

Hot Isostatic Pressing - HIP'22
Materials Research Proceedings 38 (2023) 17-21

Materials Research Forum LLC
https://doi.org/10.21741/9781644902837-3

Manufacturing of Compact Heat Exchangers by Hot Isostatic Pressing

Emmanuel Rigal[1,a*], Isabelle Moro-Le Gall[1,b], Matthieu Maunay[1,c],
Sébastien Chomette[1,d], Lionel Cachon[2,e], Sébastien Vincent[2,f]

[1]Univ. Grenoble Alpes, CEA, Liten, DTCH, Grenoble, France

[2]CEA DES IRESNE, DTN, Cadarache, Saint-Paul-Lez-Durance, France

[a]emmanuel.rigal@cea.fr, [b]isabelle.moro@cea.fr, [c]matthieu.maunay@cea.fr,
[d]sebastien.chomette@cea.fr, [e]lionel.cachon@cea.fr, [f]sebastien.vincent@cea.fr

Keywords: Hot Isostatic Pressing, Diffusion Welding, Compact Heat Exchangers, Modelling

Abstract. Compact plate heat exchangers (CPHE) are made of millimetric grooved plates stacked and joined together. Among joining processes, diffusion welding is the only one that allows joining the core of the modules without filler material (e.g., braze). However, several challenges must be met to achieve such components, including management of deformation of channels, soundness of joints, non destructive controllability and so on. Part of the achievements made in the frame of the development of the sodium-gas heat exchanger modules of the ASTRID prototype of 4th generation nuclear reactors are described in this presentation, focusing on process modeling and experimental validation with 316L steel.

Introduction

Diffusion welding (DFW) is a solid state welding process where pressure is applied at elevated temperature without macroscopic deformation nor relative motion of pieces [1]. DFW can be achieved by HIP or by uniaxial pressing, with significant differences due to the more uniform character of gas pressure compared to compression via rigid plates. Uniaxial pressing is often used for the manufacturing of CPHE, however, large HIP vessels are available worldwide, instead uniaxial presses are scarce and limited in size. The size of the DFW manufacturing equipment is of importance when dealing with CPHE because large modules are needed to benefit at best of compactness (i.e., a heat exchangers made of a great plurality of small modules are less compact), particularly for nuclear applications, which deal with high thermal power.

Some of the achievements made by CEA about manufacturing of CPHE by HIP-DFW are reported in the following. This work has been made in the frame of the development of the sodium – gas heat exchangers for the 4th generation nuclear reactor ASTRID [2] which is now abandoned, however many other advanced concepts are still under consideration [3], as well as Small Modular Reactors and non nuclear applications for hydrogen management, solar receivers and so on.

Issues

Using HIP DFW to manufacture parts with embedded cooling channels is quite common since long, see for example fusion reactor applications (ITER and beyond, [4], [5]) or applications in the field of molds and tools. Cooling channels are mainly achieved using tubes or additively manufactured parts. Seal welding channels ends to the canister allows to HIP DFW the stack of pieces at high pressure, because the pressurizing gas enters the cooling circuit, thereby preventing its collapse. Alternatively, grooved parts are encapsulated under vacuum and pre-welded using low pressure to avoid excessive deformation, then the assembly is consolidated using HIP. This last solution obviously prevails for CPHE. As a consequence, the *control of the deformation of the grooves* during the low pressure DFW step is an important issue.

Hot Isostatic Pressing - HIP'22
Materials Research Proceedings 38 (2023) 17-21

Materials Research Forum LLC
https://doi.org/10.21741/9781644902837-3

Decreasing pressure to decrease deformation presents risks in terms of weld soundness. The achievement of *high weld quality* is a second issue.

The choice of a material grade with suitable "diffusion weldability" is the third issue to be solved [6]. The fourth issue is the control of the final grain size. This article deals solely with the two first issues.

Deformation

Figure 1 shows the case of a pair of plates submitted to isostatic pressure. At edges, far from the grooves, the welding stress (that acts perpendicularly to the interface) equals the pressure. At the ribs, it is several folds higher, depending on the dimensions. Two kinds of deformations can be observed: (i) excessive rib compression leading to a barrel shape, (ii) depression of the arch when the value of t is too small. Both types lead to a reduction of h.

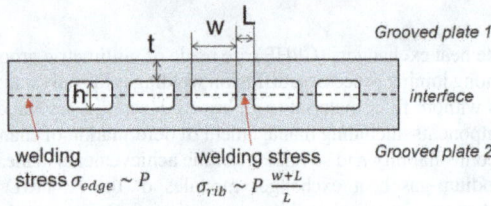

Figure 1: Pair of plates submitted to an isostatic pressure (schematic).

Figure 2 shows experimental results which illustrate deformation. It is seen for example that for this geometry at 60bar, L=2mm is required to avoid shortening of the ribs.

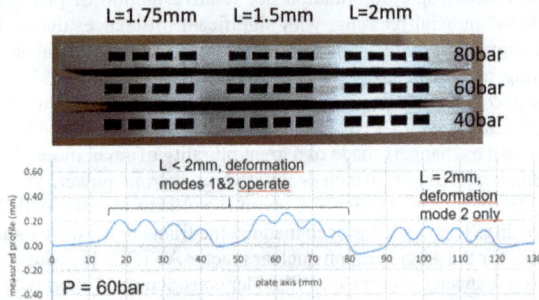

Figure 2: Left: deformation of channels, for 3 pressure values, in pairs of HIP DFWed 316L plates exhibiting 3 series of channels (w=5.75mm, h=3.5mm, t=1.75mm) with different values of L (1.5, 1.75 and 2mm). Right: surface profile showing deformation at 60bar.

The deformation of a stack of plates is more complex because channels tend to shield the welding stress, which distribute unevenly in the stack. This motivates the need for modelling, which is a powerful tool to optimize the HIP cycles and the CPHE geometry. FE modelling has been developed at CEA and has proven to give reasonably accurate results, after optimization of the constitutive law describing the high temperature behaviour of the material, Fig. 3 [7].

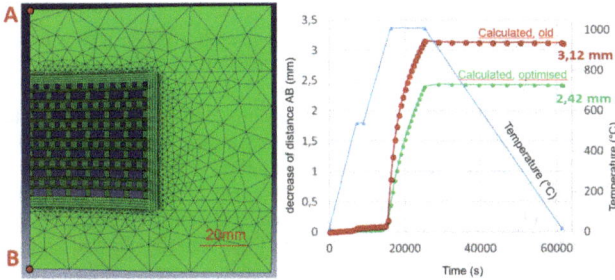

Figure 3: Left: mesh for FE modelling a stack of grooved 316L plates, Right: calculated deformation during the HIP cycle. The experimental value of AB decrease was 2.3mm.

Weld quality

Defects at DFW interfaces can be classified as (1) lack of densification (gaps, pores) and (2) microstructural imperfections (inclusions, contamination particles, uncrossed interface). The latter are not specific to CPHE manufacturing and will not be discussed here. As explained before, the process relies on a combination of low pressure DFW with encapsulated channels and high pressure HIP with open channels. Figure 4 relates to a stack of 4mm thick 316L plates. Interface characteristics are greatly improved by the HIP treatment: pores disappear and zones where the grain boundaries do not cross the interface are greatly reduced. After low pressure DFW, the joints exhibit rupture at one interface with reduced tensile strength and elongation as well as very small impact toughness compared to base material. The mechanical properties of joints largely improve after high pressure HIP, as shown in Table 1.

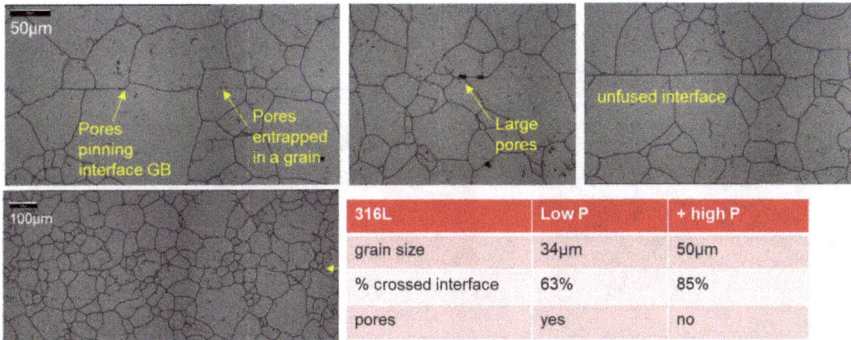

316L	Low P	+ high P
grain size	34μm	50μm
% crossed interface	63%	85%
pores	yes	no

Figure 4: Top: microstructure of 316L plates DFWed under low pressure. Bottom left: microstructure after high pressure HIP. Bottom right: interface characteristics.

Table 1: Tensile and impact toughness of DFWed 316L plate joints. (*) notch at an interface.

Cond.	Rp0.2% (MPa)	Rm (MPa)	El. (%)	RA (%)	KV (J)*
Low P	220	471 ± 16	24.3 ± 3.5	20.5 ± 3	28-30
+ high P	212	550 ±0.5	83.1 ± 3	62.5 ± 6.5	250-280
Specs.	> 220	> 525	> 45	-	-

Achievements

The ASTRID sodium-gas heat exchanger has a 1500MW thermal power. It involves a flow rate of 7200kg/s nitrogen (180bar pressure) and 6400kg/s Na. Na enters the CPHE at 530°C and leaves it at 345°C (310°C and 515°C for nitrogen).

In the course of the development program, four 40kW mock ups made of 316L grooved plates were manufactured (Fig. 5) and tested. The tests were conducted using a specifically designed loop at CEA/Cadarache. Test involved the injection of 0.2kg/s Na at 530°C and 0.2kg/s nitrogen at 310°C, 80bar (counterflow). One mock up was tested over a one year period, during which it withstood about 800 thermal cycles (120°C/min) and steady state creep at 550°C for 300h, without any signs of heat exchange degradation. Accordingly, expertise of the material showed no joint degradation, however nitride precipitation was noticed, which resulted in hardening of the material and moderate ductility loss (Fig. 6).

Figure 5: 40kW 316L mockups (1.4m long)
- *6 rows of 12 Na channels, 2x2mm*
- *6 rows of 24 gas channels, 3.5x5.75mm*

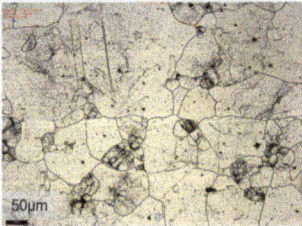

Condition	Rp0.2% (MPa)	Rm (MPa)	El. (%)	RA (%)
Before testing	219	557	86	65
After testing	230	557	69	71
specs	> 220	> 525	> 45	-

Figure 6: DFWed 316L after loop testing, showing no degradation but nitride precipitation and hardening

Summary

Compact Plate Heat Exchangers, made of grooved plates, can be diffusion welded by HIP using a combination of a low pressure cycle (vacuumed channels) and a high pressure cycle (open channels) for consolidation. The deformation of the stack can be calculated with reasonable accuracy thanks to finite element modeling. Manufacturing strategies have been developed and applied to test mockups which did not show degradation after loop testing under relevant conditions.

References

[1] Standard Welding Terms and Definitions, AWS standard A3.0M/A3.0:2010, American Welding Society, 2010

[2] L. Cachon, F. Vitillo, C. Garnier, X. Jeanningros, E. Rigal, F. Le Bourdais, S. Madeleine, O. Gastaldi, G. Laffont, Status of the Sodium Gas Heat Exchanger (SGHE) development for the Nitrogen Power Conversion System planned for the ASTRID SFR prototype, Proceedings of ICAPP 2015, May 03-06, 2015 - Nice (France), Paper 15362

[3] X. Li, R. Le Pierres, S. J. Dewson, Heat Exchangers for the Next Generation of Nuclear Reactors, Proceedings of ICAPP '06 Reno, NV USA, June 4-8, 2006, Paper 6105.

[4] P. Lorenzetto, W. Daenner, C. Boudot, P. Bucci, K. Ioki, S. Tahtinen, Manufacture of primary first wall panel prototypes for the ITER blanket shield modules, 20th symposium on fusion engineering, 14-17 oct 2003

[5] A. Cardella, E. Rigal, L. Bedel, Ph. Bucci, J. Fiek, L. Forest, L.V. Boccaccini, E. Diegele, L. Giancarli, S. Hermsmeyer, G. Janeschitz, R. Laesser, A. Li Puma, J.D. Lulewicz, A. Moeslang, Y. Poitevin, E. Rabaglino, The manufacturing technologies of the European breeding blankets, J. Nucl. Mat. 329–333 (2004) 133–140

[6] N. Bouquet, E. Rigal, S. Chomette, Influence of industrial sheet surface finish on interface formation during HIP-bonding of austenitic stainless steel, Int. Conf. HIP, Stockholm, Sweden, 2014

[7] M. Maunay, PhD thesis, university Grenoble Alpes, 2018 (in French)

Dissolvable HIP Space-Holders Enabling more Cost Effective and Sustainable Manufacture of Hydrogen Electrolyzers

Iain Berment-Parr[1,a*], Owen Larkin[1], Bea Howarth[1], Kieran Bullivant[1]

[1]The Manufacturing Technology Centre, High Value Manufacturing Catapult Network, UK

[a] iain.berment-parr@the-mtc.org

Keywords: Powder Hot Isostatic Pressing, Shape Holding Insert, Porous Structure, Titanium, Salt, Sodium Chloride, Sodium Aluminate, Hydrogen, Electrolyzer, Fuel Cell

Abstract. Polymer Electrolyte Membrane (PEM) electrolyzers are a key to the future of global hydrogen production. However, current systems rely on titanium components manufactured through energy and resource intensive processes which make up a large proportion of the overall capital cost of an electrolyzer stack. In this work circular economy principles have been applied to investigate net shape powder manufacturing routes for these titanium plate and porous film components. Approaches include (1) direct HIP of waste stream materials such as un-melted titanium sponge fines or subtractive machining swarf (2) net shape manufacturing of complex geometries using innovative dissolvable salt space holding inserts (3) in-situ nitriding methods (4) streamlining a large number of processing stages within the existing supply chain. In order to assess the environmental impact of the proposed manufacturing routes an embodied carbon analysis was conducted comparing the emissions potentially generated via this powder process versus the traditional supply chain.

Introduction

Electrolysis is a leading hydrogen production pathway to achieve the US Department of Energy 2021 Hydrogen Energy Earthshot goal of reducing the cost of clean hydrogen by 80% to $1 per 1 kilogram in 1 decade ("1 1 1") [1]. Hydrogen produced via electrolysis can result in net-zero greenhouse gas emissions, depending on the source of the electricity used. Polymer Electrolyte Membrane (PEM) electrolyzer systems extract high purity hydrogen gas directly from water, and are widely seen as one of the most commercially viable technologies to ramp up the production of "green" hydrogen derived from renewable electricity [2].

An extensive technical assessment of suitable materials capable of enabling more economical hydrogen generation has concluded that titanium is the only commercially viable base material capable of meeting the expected service life of PEM electrolysis bipolar plate and porous transport layer components, particularly at the anode which experiences high electrochemical potentials and acidity under dynamic loading conditions [3]. As a result, titanium is currently a major contributor to the overall cost of PEM electrolyzer equipment [4]. Net shape powder metallurgy is therefore an interesting area of research with the potential to reduce overall material consumption by minimizing waste. Utilization of recycled or minimally processed titanium powder sources within a circular economy framework would also help ensure the electrolyzer stack has the lowest possible embodied carbon emissions. To that end, this paper presents a set of Hot Isostatic Pressing (HIP) feasibility trials aiming to showcase an innovative consolidation route, using dissolvable space holding inserts and unconventional powder feedstocks. This method could feasibly be used to fabricate highly intricate PEM bipolar plates with greater design freedom, allowing drastic changes such as integrating porous transport layers directly into the bipolar plate. Such a design may also improve operational efficiency and make inroads into reducing the Levelized Cost of Hydrogen (LCOH) production.

Materials Research Forum LLC
https://doi.org/10.21741/9781644902837-4

Notably, the unconventional HIPing concepts developed here are likely to appeal for many other possible end-uses beyond the hydrogen value chain. Any application requiring complex powder metallurgy parts to be made at low cost and with minimal embodied carbon emissions could benefit, particularly where internal cavities or tailored porosity are required.

Dissolvable Space Holding Inserts

Powder Hot Isostatic Pressing (Powder-HIP) involves the application of sufficient heat and isostatic fluid pressure to deform an evacuated canister holding loose powder. This enables solid state inter-diffusion of the powder particles over a sufficient period of time to densify them into a solid object. This can lead to very low levels of porosity defects and an excellent balance of microstructural and mechanical properties, often comparable or better than that of cast and wrought engineering alloys.

One well documented drawback for Powder-HIP of intricate metallic components is the need to fabricate a complex canister via a welding/joining or forming process, then remove it through a hazardous and slow chemical leaching process. These processes often make costs and production timescales prohibitive for high throughput components.

In an attempt to address this fact, the authors have created an innovative approach that uses pre-formed dissolvable salt space-holding inserts, strategically positioned to guide the consolidation of powder into a relatively tightly tolerance geometry after full densification. These allow the use of relatively low cost cylindrical or cuboidal steel HIP canisters. The post-consolidation steps are also simplified by using this method as such simple canisters can be mechanically cut and peeled away rather than requiring chemical leaching in an acid bath. This process is aided by the use of a simple diffusion barrier (such as graphite sheet or a boron nitride coating) between the canister and the contents. The salt space-holding sections within the densified part can be removed by gradual dissolution in water (as long as an access route for flowing water is maintained). This only leaves a non-hazardous brine to dispose of or re-use. If the salt insert has suitable thermo-mechanical properties it is envisaged that a true net shape component could be fabricated with no subsequent subtractive machining needed.

High purity NaCl "table salt" crystals have a melting point of 801°C. Their tensile proof strength has been shown to reduce considerably above 300°C [5], but compressive strength is likely to be maintained at much higher temperatures, approaching the melting point under near isostatic loading. NaCl is therefore likely to be compatible with the consolidation of powder materials that are readily HIP consolidated below around 750°C.

Other soluble ionic compounds do exist with far higher melting points, most notably Sodium Aluminate ($NaAl_2O_3$) which has a melting point of 1650°C, meaning it is likely to be compatible with the high temperature and pressure Powder-HIP consolidation required for most engineering alloys.

Insert Preparation

The manufacturing concepts explored in this work include the formation of both solid salt inserts, designed to hold a geometric space during Powder-HIP and be dissolved completely post-HIP, as well as the formation of porous structures. The use of NaCl salt as a space-holding insert for manufacture of titanium porous structures and foams has been widely investigated in the biomedical field, primarily through press and sinter [6] or hot pressing technologies [7]. The authors therefore have high confidence than similar structures can be created using the HIP method outlined in this work. The key requirement for fully dissolving the salt portion is that a fully interconnected network of both titanium and salt is formed during HIP. The best way to achieve this is the carefully control the particle size (normally the titanium powder is much finer than the salt particles) and use a near 50:50 blending ratio by volume [7].

There are a multitude of methods that could be used to form pure NaCl into a pre-shaped insert with relatively high density. The methods trialled were:

1. Single crystal NaCl cast by the Kyropoulos method at a commercial vendor, and purchased in the form of polished transparent discs, termed "optical windows". These were utilized in the as-received form (Fig. 1a), but internal work has proven that they can be subtractively machined into a more complex geometry.
2. Cold Isostatic Pressing (CIP) within an additively manufactured flexible polymer mold (Fig. 1b) was used to fabricate a green compact insert from mechanically ground NaCl salt granules at 400-600MPa with the required geometry (Fig. 1c). The cold pressing method described above opened up additional opportunities to create bi material titanium/NaCl inserts, by either building up different layers of each material, or fully blending materials at a predetermined ratio within a CIP mold (Fig. 1d).
3. An additional method was employed for $NaAl_2O_3$, which was not pre-formed into a solid insert prior to HIP. Instead $NaAl_2O_3$ and Ti-6Al-4V were simply weighed out into the appropriate ratio, then vibratory blended before being layered manually directly within the HIP canister. $NaAl_2O_3$ salt is significantly more hazardous than NaCl table salt, and as a precaution all powder handling was undertaken in a dry enclosed glovebox environment in order to avoid any moisture reacting with the finely divided salt particles which are easily airborne.

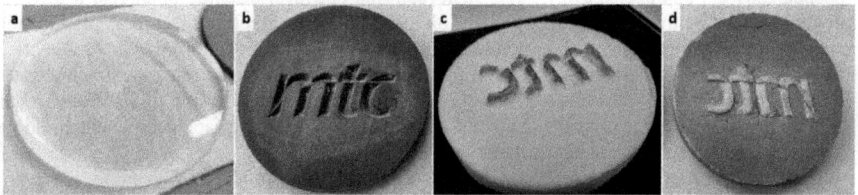

Figure 1: a) Commercially available single crystal NaCl "optical window", b) flexible polymer CIP mold with lettering, c) NaCl disc CIP compacted from granules at 400MPa (lettering filled with Ti-64 + NaCl blend), d) Ti-64 + NaCl blend disc CIP compacted from granules at 400MPa (lettering filled with pure NaCl).

Low Embodied Carbon Powder Materials
In order to enhance the circular economy credentials of HIP components, this work has also evaluated powder forms of titanium with potentially lower embodied carbon in their production than conventional gas or plasma atomization processes. These included those listed below and shown in Fig. 2:

1. Commercial purity titanium (CP-Ti) and Ti-6Al-4V obtained through the solid-state hydride-de-hydride (HDH) pulverization process at an external vendor were used.
2. CP-Ti in the form of Kroll extracted titanium "sponge dropout waste" was obtained from a commercial titanium supplier, sieved to a sub-75μm fraction, and directly HIPed.
3. CP-Ti ingot and billet machining chip was also obtained from a commercial titanium supplier and attempts were made to mechanically ball-mill it down to a usable powder fraction, but the resulting material was not suitable due to excessive contamination.

It is recognized that all of these re-processed/waste materials will have significantly higher interstitial element content, most notably oxygen and carbon, however, prior work in the literature

Materials Research Forum LLC
https://doi.org/10.21741/9781644902837-4

has proven titanium's ability to vacuum sinter and diffusion bond with good mechanical bond strength despite the presence of such impurities. For example, angular 144 177µm HDH CP-Ti powder containing 1500ppm oxygen had been successfully HIP consolidated to 98.1% density after a 90 minute dwell at 700°C and 34MPa, with notably high rates of densification in the early stages relative to more spherical powders due to higher stresses at angular interfaces [8].

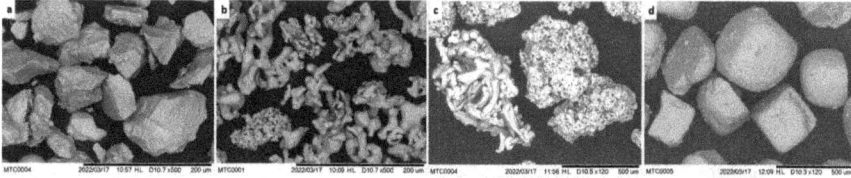

Figure 2: SEM powder analysis images of a) -75µm sieved Ti-6Al-4V HDH (x500), b) -75µm sieved CP-Ti sponge (x500), c) +75µm sieved CP-Ti sponge as received (x120), d) NaCl granules as received (x120).

HIP Conditions

All titanium powder and NaCl salt layers were manually spread within 65mm outer diameter cylindrical mild steel canisters under an air environment. NaAl$_2$O$_3$ blended powders were spread under an enclosed argon glovebox environment as fine salt particles easily became airborne. TIG welding was employed under argon back purge and leak checking confirmed a suitable seal had been achieved. Canisters were evacuated and crimped shortly before HIP. All HIP cycles included a 15°C/min ramp up in temperature and after the hold period natural cooling to 200°C was employed before forced cooling to room temperature and pressure.

The relatively soft commercial purity titanium grade 2 (CP-Ti) has been chosen as a material capable of full Powder-HIP consolidation under the low temperatures and pressures compatible with NaCl shape-holding inserts. Conditions of 750°C and 35MPa for 4 hours were targeted based on the work of Lograsso et al [8].

Conventionally, higher strength titanium alloys such as grade 5 (Ti-6Al-4V) are often HIP consolidated at temperatures of around 900-950°C and pressures of ~100MPa for ~2 hours [9]. However, due to the requirements of the NaCl space-holding inserts, in these feasibility trials Ti-6Al-4V powder material was HIPed using the same low temperature and pressure conditions as for CP-Ti.

A subsequent set of HIP trials investigated the potential use of sodium aluminate salt (NaAl$_2$O$_3$) for more conventional high temperature Powder-HIP of Ti-6Al-4V at 920°C and 103MPa for 2 hours. To the authors knowledge no assessment of NaAl$_2$O$_3$ as a space holding insert during Powder-HIP of titanium or any similar powder alloys has been reported in the open literature.

Post-HIP Characterization

Once HIP cycles were completed the canisters were cross-sectioned without the use of cutting fluid in order to avoid dissolving the salt inserts. Dry grinding and polishing attempted to produce a clean surface for imaging, but the result was less uniform than typical fluid-based grinding and polishing material preparation techniques.

Figure 3 summarizes the space-holding and porous features achieved within canister 1, which included the various NaCl salt and blended salt + titanium layers labelled. As can be seen, the canister 1 trial has successfully proven a number of concepts:

1. Both CP-Ti and Ti-6Al-4V HDH can be Powder-HIP consolidated to high density and bonded within the same canister under 750°C / 35MPa / 4 hour processing conditions.

Hot Isostatic Pressing - HIP'22   Materials Research Forum LLC
Materials Research Proceedings 38 (2023) 22-28   https://doi.org/10.21741/9781644902837-4

Notably, other canister trials have also proven CP-Ti sponge will successfully consolidate at these conditions.

2. Both single crystal NaCl and CIP pre-formed NaCl inserts have successfully held a free space within the HIP part. In regions of inconsistent strain (most notably the domed base of the canister where the steel was not as thick) show that the NaCl inserts do deform preferentially compared to the consolidating Ti-6Al-4V powder, which is perhaps unsurprising given their differing compressive strength at the processing temperature.

3. A porous structure has been formed using blended NaCl salt granules and Ti-6Al-4V powder, and this has successfully bonded to a consolidating CP-Ti plate as would be required in the fabrication of a PEM hydrogen electrolyzer bipolar plate. More work is needed to control the blending process and produce a homogeneous and interconnected porous network, but this sample clearly indicates that the method would be viable with refinement.

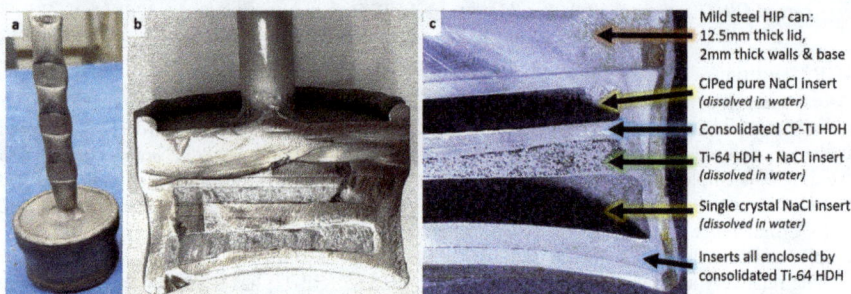

Figure 3: a) as-HIPed canister 1, b) dry cross-sectioned canister 1, c) labelled detail within canister 1 cross section after water dissolution of the NaCl salt space-holding portions.

Canister 2 (shown cross-sectioned in Fig. 4) contained a gradual transition in blended $NaAl_2O_3$ particles and Ti-6Al-4V HDH. This has successfully proven the space-holding capabilities of this ionic compound under a conventional Ti-6Al-4V HIP cycle, with both the pure salt section being easily removed to form an open cavity, and the porous structure being well defined. It was noted that post-sectioning, on exposure to air on a humid day, the $NaAl_2O_3$ salt did begin to spontaneously dissolve in air, highlighting some potential material handling challenges were this process ever to be up-scaled to a production environment.

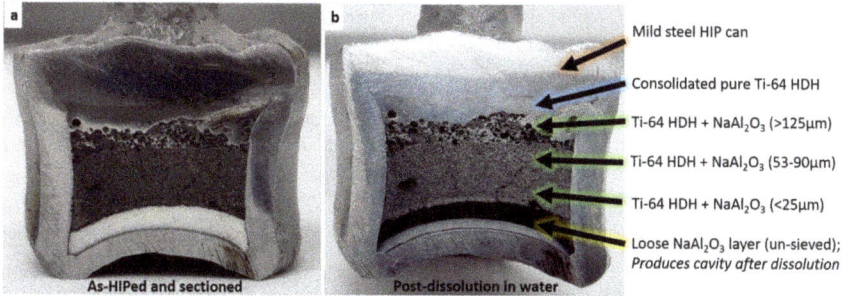

Figure 4: Cross-section of Canister 2 showing a) as-HIPed cross section with NaAl$_2$O$_3$ space-holders visible, and b) graded porous structure revealed post-dissolution in water.

Embodied Carbon Analysis of HIP Concept

A detailed embodied carbon analysis has been undertaken to compare against conventional titanium manufacturing routes [10]. A baseline emissions value has been estimated for a PEM electrolyzer stack containing forty individual cells, each consisting of one bipolar plate and two porous layers, manufactured using traditional commercial methods.

Estimates were then made for the total embodied carbon of various alternative titanium processing and bipolar plate forming concepts. The output from this study is shown graphically in Fig. 5. This indicates that significant reductions to the manufacturing carbon footprint can be achieved through use of HIP consolidation of recycled end-of-life titanium parts HDH crushed into usable powder (noting calculation does not take account of any carbon emissions prior to recycling).

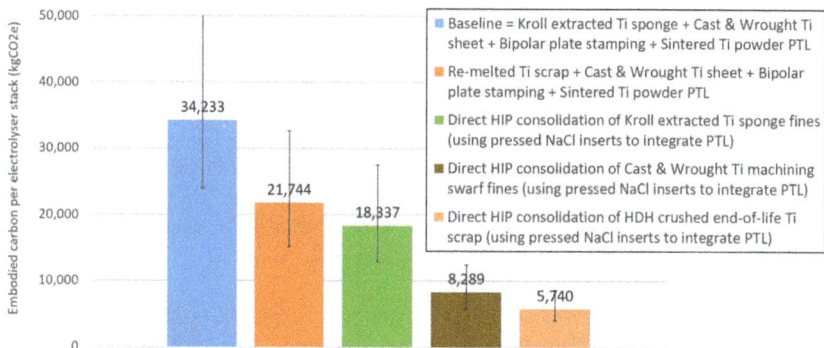

Figure 5: Estimated total embodied carbon for all titanium within an electrolyzer stack (based upon a 40-cell stack containing 260Kg of commercial purity titanium)

Conclusions

1. The feasibility studies presented highlight a new opportunity to utilize HIP to consolidate powder with intricate internal or surface features, as well as tailored through porosity, using dissolvable space holders. Space holders fabricated from NaCl table salt have been shown to be suitable for HIP consolidation at 750°C and 35MPa, whilst NaAl2O3 salt has been shown to be compatible with more conventional titanium HIP consolidation at 920°C and 103MPa.

2. A possible new route to net-shape manufacture intricate titanium PEM electrolyzer bipolar plates via HIP has been demonstrated. The potential design and performance benefits of this method have been highlighted, particularly regarding the ability to integrate porous transport layers into the plate. Notably this concept could be applied to reimagine manufacture of many other complex components beyond those used in the hydrogen energy sector.
3. An embodied carbon analysis exercise has indicated that significant circular economy benefit may be realized versus established manufacturing methods.

Acknowledgments
The authors thank Titanium Metals Corporation (PCC TIMET) for provision of titanium materials.

References
[1] B. Pivovar, Low Temperature Electrolyzer Technology, Electrolysis Panel Session Presentation at the U.S. Department of Energy (DOE) Hydrogen Shot Summit, August 31-September 1, 2021.

[2] E. Taibi et al, International Renewable Energy Agency, Green Hydrogen Cost Reduction - Scaling up Electrolysers to Meet the 1.5°C Climate Goal (2020): Published online at www.irena.org/publications.

[3] Ole Edvard Kongstein et al, Alternative Energies and Atomic Energy Commission (CEA), NEXPEL Project WP5 "Porous current collectors and materials for bipolar plate" bibliographic review, 2010.

[4] Luca Bertuccioli et al, E4tech Sàrl and Element Energy, Study on development of water electrolysis in the EU, Approved for publication by the Fuel Cells and Hydrogen Joint Undertaking, 2014.

[5] P.L. Pratt and R.P. Harrison, Dislocation Mobility in Ionic Crystals, Discussions of The Faraday Society No. 38, Aberdeen University Press, page 211-217, 1964. https://doi.org/10.1039/df9643800211

[6] S. Lascano et al, Porous Titanium for Biomedical Applications: Evaluation of the Conventional Powder Metallurgy Frontier and Space-Holder Technique, Journal of Applied Sciences, 9, 982, 2019. https://doi.org/10.3390/app9050982

[7] B Ye and D. C. Dunand, Titanium foams produced by solid-state replication of NaCl powders, Materials Science and Engineering A, issue 528, page 691-697, 2010. https://doi.org/10.1016/j.msea.2010.09.054

[8] B.K Lograsso et al, Densification of titanium powder during hot isostatic pressing, Metallurgical Transactions A, Volume 19A, 1767-1773, 1988. https://doi.org/10.1007/BF02645145

[9] U.M. Attia, HIPing of Pd-doped titanium components: A study of mechanical and corrosion properties, Proceedings of the 11th International Conference of Hot Isostatic Pressing, 2014.

[10] Full source references for the reported embodied carbon analysis are available upon request.

Hot Isostatic Pressing - HIP'22
Materials Research Proceedings 38 (2023) 29-34

Materials Research Forum LLC
https://doi.org/10.21741/9781644902837-5

The Pathology of PM HIP Duplex Stainless Steels

Tomas Berglund[1,a *], Björn-Olof Bengtsson[2,b], Jan-Olof Nilsson[3,c]

[1]MTC Powder Solutions, Surahammar, Sweden

[2]IKnowPowder AB, Stockholm, Sweden

[3]JON Materials Consulting, Arvika, Sweden

[a]tomas.berglund@mtcpowdersolutions.com, [b]bjornolof@iknowpowder.se,
[c]janolof.jedviknilsson@gmail.com

Keywords: Powder Metallurgy, PM, Hot Isostatic Pressing, HIP, Duplex Stainless Steel, DSS, Super Duplex Stainless Steel, SDSS

Abstract. Tests were conducted to simulate possible issues in manufacturing of Powder Metallurgical HIPed Duplex Stainless Steels. Root causes, and consequences are analyzed and discussed from a manufacturing, metallurgical and properties point of view. The results highlight the importance of material understanding and good process control when manufacturing these alloys. While some issues are unique to PM HIP material, many of them can also be found in conventional wrought materials e.g., sigma phase and nitride precipitation. In addition, the findings in this study puts into question limitations stated in some specifications for this process and alloys. The findings show the importance of staying within these limitations but also show that some aspects are not as critical. The majority of these specifications are based on forging specifications that might result in unnecessary limitations on the PM HIP process and materials. This while not necessarily ensuring material quality or possibly limiting material use.

Introduction

Duplex Stainless Steels are characterized by high mechanical strength combined with excellent resistance to stress corrosion cracking, pitting and crevice corrosion and general corrosion. In applications for the Subsea Oil & Gas Industry and chemical industry the demands on the material are getting tougher as operating pressures and temperatures are increasing combined with an increasing demand for material integrity. This is pushing the limits on the manufacturability of conventional forged materials and the industry is moving more and more to PM HIP materials as conventional forging cannot meet the tougher requirements.

Duplex Stainless Steels contain an approximate 50/50 mixture of ferrite (α) and austenite (γ) and has a fairly complex metallurgy that if not processed properly can cause a number of issues. Many of these issues can be eliminated by the use of PM HIP, e.g., segregation caused by insufficient hot working and other microstructural defects associated with forgings. The major limiting factor for PM HIP DSS is shared with the conventional materials, the precipitation of unwanted phases during heat treatment, i.e., the formation of embrittling intermetallic and nitride phases during water quenching following the solution annealing. Even small amounts of intermetallic and/or nitride phase may affect the impact toughness and/or corrosion resistance adversely for DSS and SDSS components, although this is not always the case, especially for intermetallic phase. The formation of unwanted phases is most often a consequence of poor heat treatment process, possibly in combination with a large wall thickness. It can also be a consequence of poor chemistry caused by an overly wide alloy composition range. To find the detrimental precipitates, Light Optical Microscopy is used as the standard test method in manufacturing testing according to most customer specifications. Sometimes LOM might prove insufficient as some detrimental precipitates are so small that they require more sophisticated analysis methods in order to be found.

One of the major benefits of PM HIP DSS is the fine isotropic microstructure in the material. The γ-spacing of PM HIP DSS is normally in the range of 10-15 µm and it is essentially the same throughout the component whether it is a 1kg part or a 10-ton part. If we should compare this point to large forgings, they are rarely below 30µm without very special processing. In the Oil & Gas industry the small γ-spacing provides a major benefit with an increased resistance to Hydrogen Induced Stress Cracking [1]. In the Urea industry the fine γ-spacing improves the corrosion resistance to ammonium carbamate as the negative effects of the preferential corrosion of the γ-phase can be limited [2].

Unlike conventional materials the level of oxygen content in PM HIPed components is extremely important as it has a significant influence on properties, especially impact toughness [3, 4, 5]. In this study the effect of oxygen has been compensated for to examine other factors when the measured oxygen content cannot explain a deviation from the expected properties.

Experimental

The test material in this study has been manufactured to simulate issues that can and do appear in the manufacturing of PM HIP SDSS components. All the tested material is in a thickness range that in normal manufacturing does not result in issues with properties and/or microstructure. The alloys used in this study is APM2327 (UNS S32505) and APM2329 grade (UNS S32906). The nominal chemistry for each alloy can be seen Table 1.

Table 1. Nominal chemical composition of alloys in this study.

	C	Cr	Mo	Ni	Cu	N	Fe
APM2327	<0,03	26	3	7	2	0,27	Bal.
APM2329	<0,03	29	2,3	7	<0,8	0,35	Bal.

Powders used in this study was manufactured using nitrogen gas atomization. The particle size is -500µm with a d_{50} of 100-120 µm. The materials have been HIPed at 1150°C, 1000 bar with 1-3 hours dwell time. APM2327 (UNS S32505) solution annealed at 1070°C and APM2329 (UNS S32906) at 1060°C, both followed by quenching in cold water. The material thickness and quench rate has been varied to simulate process variations that can be seen in regular manufacturing.

Mechanical testing has been performed according to ASTM A370. Microstructural analysis has been performed using LOM on specimens etched with 10% oxalic acid in a first step to reveal precipitates like carbide and nitrides. In a second step the sample is electrolytically etched in 20% NaOH to reveal intermetallic phases like Sigma and Chi. Ferrite content has been measured according to ASTM E 562-11 and γ-spacing has been measured according to DNV-RP-F112.

SEM/EDS has been used for more in depth-analysis of precipitates. EBSD has been used for evaluating phase distribution, phase identification and to measure the amount of sigma phase.

Results

Intermetallic precipitation. There are a number of intermetallic phases that can form in DSS e.g., Sigma (σ-phase), Chi, and R phase. In general, it is only the σ-phase that has been found to be the limiting for the more common PM HIP SDSS. All these intermetallic phases affect the properties, but σ-phase is the dominating phase, at least after longer aging times. σ-phase is a Mo and Cr rich, hard embrittling precipitate that nucleate and grow primarily in α/α and α/γ phase boundaries in the approximate temperature interval of 600 - 1000°C [6, 7]. At 900°C it can take as little as 2 minutes for the α in a 25Cr SDSS to transform to σ-phase. Due to the rapid precipitation this is also the main limiting factor as to how thick section can be manufactured as the cooling rate after

Hot Isostatic Pressing - HIP'22   Materials Research Forum LLC
Materials Research Proceedings 38 (2023) 29-34   https://doi.org/10.21741/9781644902837-5

HT must be fast enough to avoid the precipitation. Table 2 shows the typical thickness range for a selection of DSS manufactured by PM HIP, limited by precipitation of σ-phase.

Table 2. Typical maximum thickness of common PM HIP DSS.

APM2377	APM2328	APM2327	27Cr	APM2329
UNS S31803	UNS S32750	UNS S32505	UNS S32707	UNS S32906
300-350mm	100-150mm	175-225mm	50-100mm	250-300mm

Any σ-phase in the microstructure will affect the material negatively. Studies performed have shown that even small amounts, less than 0.1% of σ phase can cause a large drop in impact toughness [8]. However, for it to affect the corrosion resistance there needs to be a substantial amount of σ-phase in the microstructure, typically above 0,5% [9, 10]. Even at low amounts, the precipitates are usually clearly visible, even at moderate magnifications during microstructural investigation using SEM or LOM on etched specimens. Fig. 1 shows typical LOM micrograph of σ-phase in APM2327 SDSS. The σ phase is indicated with red arrows.

The presence of σ-phase in the material does not immediately disqualify the material as it may still have good mechanical and corrosion properties. The micrograph in Fig. 1 is from T/2 (mid-section) on a APM2327 test piece with 191 mm thickness, i.e., close to the limit for what is practically feasible with APM2327. The test piece was positioned poorly in the quench tank so water flow around it was low during quenching and consequently the cooling rate was lower than ideal. Measurements in LOM indicates the volume of σ-phase is 0,1-0,17%. Analysis with EBSD confirmed 0.1%. Despite this the material had an impact toughness of 83J at -46°C, well above the acceptance criteria of 40J. Furthermore, corrosion testing using ASTM

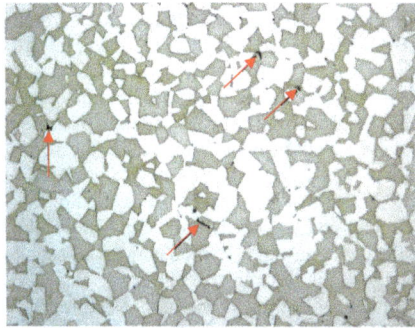

Figure 1. Typical microstructure of APM2327 containing a small amount of sigma phase.

G48 method A at 50°C showed no weight loss. Another example, a large production part with 178mm section thickness with a design causing significant restrictions to water flow in and around the part during quenching. Most of these parts had σ-phase at T/2. The impact toughenss was 65-70J at -46°C and corrosion test showed no weight loss. Tests performed at T/4 showed only 10J higher impact toughness despite containing no σ-phase at all. In all cases the tensile properties far exceeded the minimum requirements.

Tight control of the powder chemistry in combination with a well-controlled and executed heat treatment is paramount for manufacturing a material without σ-phase. If there is σ-phase in the material, it may have many causes, but the driving factor is low quench rate in the critical temperature interval. Low quench rate can be caused by many things e.g., too large part thickness, too low agitation in tank, poor part design, poor part placement HT lot, etc. In rare cases it can also be connected to poor chemistry control and wrong HT-temperature.

Other detrimental precipitates. Most HIP DSS materials specifications are derivatives of forging specifications and only dictate LOM for investigating the microstructure. While this in most cases is sufficient to confirm that the microstructure is sound, the resolution limit of LOM can mean some things may be overlooked e.g., nitrides. Nitrides not only reduces toughness, but unlike σ-phase, also corrosion resistance already at very low amounts [10]. There are two main types of

Materials Research Forum LLC
https://doi.org/10.21741/9781644902837-5

nitrides found in Duplex stainless steel, equilibrium and non-equilibrium nitrides. Non-equilibrium nitrides, or quenched nitrides form when there is insufficient time for nitrogen to diffuse from α as the solubility gets lower with reducing temperature during quenching. The nitrides are usually found as clouds of precipitates inside larger α-grains where nitrogen has a further distance to diffuse. These nitrides are less likely to form in the more common HIPed DSS grades due to the very fine microstructure which means that nitrogen has a shorter distance to diffuse out of the α. The grain size dependence has been identified by others [11] and confirmed in experiments [12].

Equilibrium nitrides can be found in the grades of PM HIP SDSS found in Table 2. Fig. 2. shows an example of APM2329 test material that in standard testing using only LOM was reported to have a very small amount of intermetallic phase (<0,1%) and no other precipitates of any kind. Despite this, the impact toughness was only 28J at -35°C at T/4 (quarter thickness) which is surprisingly low for this material at 223mm thickness. Normally the toughness for 250mm thick material is in the range of 50-90J at -35°C at T/4. When the material was studied in SEM it became evident that there were significant amounts of nitrides in the γ-α grain boundaries across the section thickness. Using ThermoCalc software it was

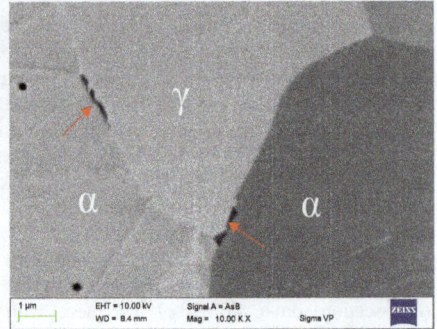

Figure 2. Micrograph of 29Cr alloy with grain boundary nitrides

concluded that an elevated Cu-content of the alloy increases the stability of the Cr-nitride phase. The Cu-content was within specification but the deviation in combination with large section thickness and unfavorable geometry for quenching meant the cooling rate was not sufficient to suppress the formation of nitrides.

Figure 3 shows another example of a APM2327 material that had lower than expected impact toughness despite only a very small amount of σ phase, <0,1% measured with EBSD. Basic LOM and SEM investigation did not provide any explanation to the low impact toughness. Only after careful sample preparation and very high magnification SEM a possible explanation came into light. 20-50 nm Cu-rich precipitates was found on γ-α grain boundaries as well as inside the α-grains, see Fig. 3. The Cu-rich precipitates have been found in several other test materials and are only found when σ is present in the material. Other studies have found that the cooling rate influences the precipitation of the Cu-rich phase and have found that Cu-precipitates could in

Figure 3. SEM micrographs on APM2327 showing Cu-rich precipitates in ferrite and austenite-ferrite grain boundaries.

fact also be found in materials without intermetallic phase, at least in other alloys than APM2327 [13]. Materials investigated in this study always contains a significant amount of these precipitates only when σ-phase is present in the APM2327 material. The exact mechanism of the formation is yet to be fully understood but is clear they can contribute to lowering the impact toughness of the material.

Phase balance and austenite spacing. As mentioned earlier, the fine austenite spacing of PM HIP DSS is often a major advantage in many applications. However, if too small the γ-spacing may have a negative effect on the impact toughness. One example of this is a production part in APM2329 that had surprisingly low impact toughness (CVN), 35J at -35°C. This despite having a good microstructure without intermetallic or any other detrimental precipitates. The reported α-content was of 49% (SD=5%) with a γ-spacing of 7,2μm (SD=0,1μm). When studying the fracture surface from the impact testing it is evident that the fracture contains, except for the initiation zone, dominantly a brittle fracture with small islands of ductile fracture between, see Fig. 4a. Figure 4b shows a EBSD mapping on a cross section of the fracture surface, just under the initiation zone. It is evident that the crack propagates through a combination of fracture of the α (red) and in the α-γ grain boundaries. The small areas of ductile fractures in the γ (blue) can also be seen. Note that severely distorted regions in the EBSD map are not mapped in Fig. 4b and appear black.

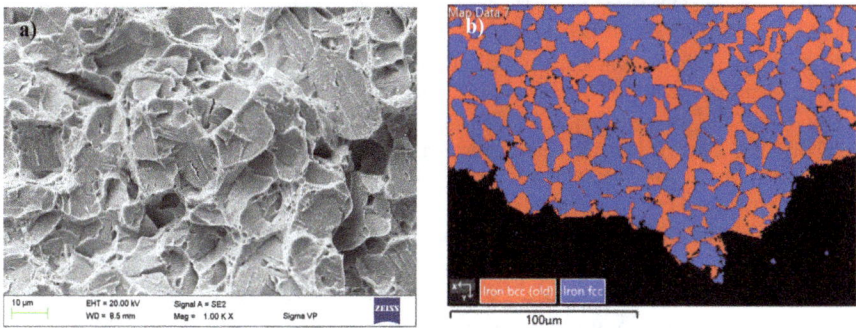

Figure 4. Micrograph of CVN specimen fracture surface (a) and EBSD mapping of fracture surface (b).

Without any obvious explanation to the poor toughness, we would propose that it is a consequence of the fine γ-spacing. With the very fine γ-spacing cracks can easily propagate as less energy is adsorbed when cracks are forced to change direction at phase boundaries. The issue can be worsened further by high α-content in the material which can essentially create a material with a α matrix further simplifying the crack propagation. Especially at lower temperature where the ferrite behaves brittle.

Conclusions
- Even if there is σ-phase present in the material, the mechanical properties can still be good
- Corrosion resistance may still be good even if 0,5% σ-phase is exceeded
- Even small amounts of σ-phase will affect impact toughness while the corrosion resistance is less sensitive to σ-phase
- In some cases, standard LOM might not be sufficient to resolve all detrimental phases that can have a negative effect on material properties
- Poor chemistry control may result in the formation of detrimental phases not seen in LOM
- Very low austenite spacing may have a negative effect on impact toughness

References
[1] U. Kivisäkk, Influence of hydrogen on corrosion and stress induced cracking in stainless steels, Doctoral Thesis, The Royal Institute of Technology, Stockholm, Sweden, (2010).

[2] T. Berglund, M. Östlund and A. Scheerder, Improved Corrosion Properties With PM HIP For Duplex Safurex®UXM? Developed For Urea Manufacturing -- Called: Safurex®UXM°, Proc. World PM 2016

[3] A. Cooper, Effect of Oxygen Content Upon the Microstructural and Mechanical Properties of Type 316L Austenitic Stainless Steel Manufactured by Hot Isostatic Pressing, Metall. Mater. Trans. A, Published online 22 June 2016. https://doi.org/10.1007/s11661-016-3612-6

[4] T. Berglund and M. Östlund, Impact Toughness for PM HIP 316L at Cryogenic Temperatures, Proc. ASME 2016 Pressure Vessels and Piping Conference, Paper No: PVP2016-64002. https://doi.org/10.1115/PVP2016-64002

[5] B-O Bengtsson, G. J. Del Corso and J. F. Scanlon, Impact Strength for PM HIP: Properties of PM HIP Stainless steels, Proc. Euro PM2012 conference, 2012, Basel, Switzerland

[6] J. O. Nilsson, Overview, Super Duplex Stainless Steels, Mat. Sci. Techn., 8 (1992) 686-700. https://doi.org/10.1179/mst.1992.8.8.685

[7] A. Wilson and J. O. Nilsson, Modelling CCT-diagrams of intermetallic phase precipitation in duplex stainless steels, Scand. J. Metall, 25 (1996) 178-185.

[8] J. O. Nilsson, P. Kangas, T. Karlsson and A. Wilson, Mechanical properties, microstructural stability, and kinetics of sigma phase formation in a super duplex stainless steel, Metall. Mater. Trans. A, 31A (2000) 35-45. https://doi.org/10.1007/s11661-000-0050-1

[9] J-O Nilsson and G. Chai, "Physical Metallurgy in duplex stainless steels", Proc. Conference on Duplex stainless steels 2010, Beaune, France, EDP Science, (2010), 268–390.

[10] L. Karlsson, Overview, Duplex Stainless Steel 2000, Associazione Italiana di Metallurgia (2000), 693-702.

[11] N. Pettersson, R. F. A. Pettersson and S. Wessman, Precipitation of Chromium Nitrides in the Super Duplex Stainless Steel 2507, Metall. Mater. Trans. A, (2015), Published online 04 January 2015. https://doi.org/10.1007/s11661-014-2718-y

[12] B. Josefsson, J. O. Nilsson and A. Wilson, Phase transformations in duplex steels and the relation between continuous cooling and isothermal heat treatment, Proc. Conf. "Duplex Stainless Steels 91", Beaune France, Les Edition de Physique, 1, 67 (1991)

[13] O. Smuk, Microstructure and properties of modern P/M super duplex stainless steels, Doctoral thesis, Royal institute of Technology, Stockholm, ISBN: 91-7283-761-6

Hot Isostatic Pressing - HIP'22
Materials Research Proceedings 38 (2023) 35-40

Materials Research Forum LLC
https://doi.org/10.21741/9781644902837-6

Application of HIP-NNS to Large Complex Products Using Super Duplex Stainless Steel Powder

Toyohito Shiokawa[1,a] *, Mitsuo Okuwaki[2,b] and Hiroshi Urakawa[3,c]

[1]Metal Technology Co., Ltd., Himeji-shi, Hyogo-ken, Japan

[2]Metal Technology Co., Ltd., Sawa-gun, Gumma-ken, Japan

[3]Shimoda Iron Work Co., Ltd., Aioi-shi, Hyogo-ken, Japan

[a]tshiokawa@kinzoku.co.jp, [b]okuwaki@kinzoku.co.jp, [c]h.urakawa@shimoda-flg.co.jp

Keywords: Near Net Shape, Super Duplex Stainless Steel, NORSOK

Abstract. In recent years, the productivity of offshore plants has been improved and the field of production from shallow water to deep sea has been expanded. Along with this, it is required that the plant equipment be upsized and maintenance-free. The conventional manufacturing method for duplex stainless steel parts such as valves for plant equipment was mainly forging. There is a problem though, it is difficult to obtain homogeneity with the physical characteristics. One solution to this problem is powder sintering using HIPed near net-shaped (HIP-NNS) products. However, near-net manufacturing requires advanced capsule design and powder filling control. In this paper, we will report the results of the HIP-NNS method when manufacturing a large-sized complex-shaped powder sintered product using the world's largest HIP device.

Introduction

In recent years, the oil and gas industry has been developing offshore oil fields, with the development moving from shallow to deep water. One of the by-products of deep-sea oil exploitation is corrosive materials such as hydrogen sulfide (H_2S) and carbon dioxide gas (CO_2). These highly corrosive by-products require components and equipment that can withstand both high pressure and corrosive fluids for a long period of time. Therefore, duplex stainless steel and super duplex stainless steel are often used. The hostile environment of deep-sea oil extraction also requires maintenance-free parts. In addition, there is a need to increase the size of parts produced in order to facilitate increased production. Conventionally, the manufacturing method for valves, etc. has been mainly done through forging. However, since it is difficult to obtain uniformity of mechanical properties such as strength and corrosion resistance at the weld at the time of assembly, a weld-less integrated shaping technique is required. Therefore, as a method to replace forging, near-net shaping by HIP is used. By utilizing this method, it can be expected that material weight and machining time will be reduced. [1] In this study, in order to manufacture large parts using MTC'S Giga-HIP, a large flange and valve body were manufactured using UNS S32505 powder, which is super duplex stainless steel. The resulting shape was then compared with the designed shape and the mechanical properties due to differences in heat treatment conditions were also investigated.

Dimension evaluation - Flange

Method

Fig. 1 shows the target shape after HIP. Based on this target shape, the capsule was designed assuming a powder filling rate determined using a pretest. In order to allow the capsule to shrink evenly, mild steel with a plate thickness of 3.2 mm was used for all sections. [2] The powder used was UNS S32505, which is a super duplex stainless steel. The powder filling was facilitated with vibration. The actual filling rate of the powder was 7% higher than the assumed rate. The HIP treatment temperature was 1150 °C [3] and the pressure was 118 MPa. Dimensional measurements were performed before and after HIP with the target shape and the actual shape compared after HIP.

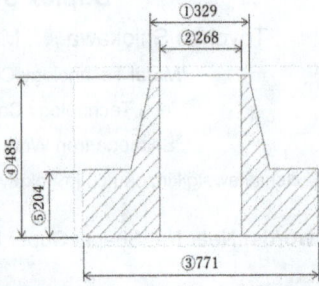

Figure 1. Target shape of Flange

Results

The dimensional measurement results are shown in Table 1. Dimensions were measured at the locations described in the dimensions shown in Fig. 1. When comparing the target dimensions with the actual dimensions after HIP, the maximum difference was 4.7%. In order to eliminate errors due to the difference in filling rates, the expected dimensions after HIP and the actual dimensions after HIP when the capsules were filled with were compared. The difference was 1.2%. Figs. 2 and 3 show the measurement results with a 3D digital analyzer. Fig. 2 compares the design shape of the capsule before HIP and the actual shape. The maximum difference between the design dimensions and the measured dimensions was 3.3 mm. Fig. 3 compares the expected shape after HIP and the actual shape after HIP calculated using the actual filling rate. The maximum error was 5.6 mm. The difference in maximum error between Figs. 2 and 3 was 2.3 mm.

Table 1. Dimension measurements results of Flange

	①	②	③	④	⑤
After HIP [mm]: A	343	280	801	508	212
Goal shape [mm]: G	329	268	771	485	204
Error [%]: E	4.3	4.5	3.9	4.7	3.9
After HIP shape AFR [mm]: A_{AFR}	341	278	798	502	211
Error AFR [%]: E_{AFR}	0.6	0.7	0.4	1.2	0.5

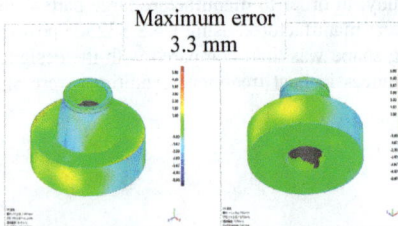

Fig. 2 Compares the design shape of the capsule before HIP and the actual shape

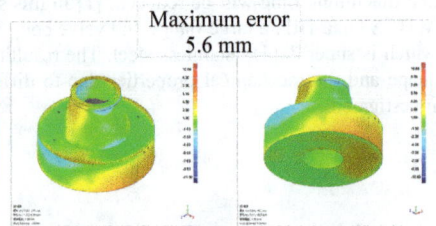

Fig. 3 Compares the expected shape after HIP and the actual shape after HIP calculated using the actual filling rate

Dimension evaluation - Valve body

Method

The target shape of the valve body is shown in Fig 4. Based on this target shape, the capsule was designed assuming a powder filling rate the same as the actual rate for the flange. The capsule was made of mild steel with a plate thickness of 3.2 mm. The powder used UNS S32505, which is a super duplex stainless steel, and the powder was filled using a vibrator. The filling rate was 3.0% lower than the designed value. As with the flange, the HIP treatment temperature was 1150 °C and the pressure was 118 MPa. Dimensional measurements were performed before and after HIP, and the target shape and the actual shape after HIP were compared.

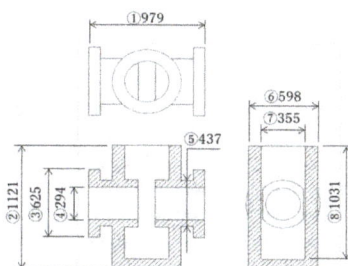

Figure 4. The target shape of Valve body

Results

The dimensional measurement results are shown in Table 2. Dimensions were measured at the locations described in the dimensions shown in Fig. 4. When comparing the target dimensions with the actual dimensions after HIP, the difference was up to 5.7%. In order to eliminate errors due to the difference in filling rates, the expected dimensions after HIP calculated by the actual filling rate and the actual dimensions after HIP were compared. The difference was 2.3%. Figs. 5 and 6 show the measurement results with a 3D digital analyzer. Fig. 5 compares the design shape of the capsule before HIP with the actual shape. The maximum difference between the design dimensions and the measured dimensions was 13.2 mm. Fig. 6 compares the expected shape after HIP and the actual shape after HIP calculated using the actual filling rate. The maximum error was 15.7 mm. The difference in maximum error between Figs. 5 and 6 was 2.5 mm.

Table 2. Dimension measurements results of Valve body

	①	②	③	④	⑤	⑥	⑦	⑧
After HIP [mm]: A	949	1110	610	294	412	571	360	1030
Goal shape [mm]: G	979	1121	625	294	437	598	355	1031
Error [%]: E	-3.1	-1.0	-2.4	0	-5.7	-4.5	1.4	-0.1
After HIP shape AFR [mm]: AAFR	943	1105	604	291	406	574	352	1013
Error AFR [%]: E_{AFR}	0.6	0.5	1.0	1.0	1.5	-0.5	2.3	1.7

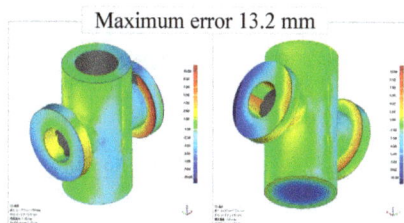

Maximum error 13.2 mm

Maximum error 15.7 mm

Fig. 5 Compares the design shape of the capsule before HIP with the actual shape

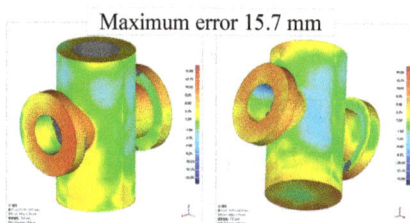

Fig. 6 Compares the expected shape after HIP and the actual shape after HIP calculated using the actual filling rate

Discussions

The reason why the target dimensions of the flange and the actual dimensions after HIP were off by 4.7% is thought to be that the capsule shrank less than expected because the powder was filled 7% more than the expected filling rate. The reason why the target dimensions of the valve body and the actual dimensions after HIP were off by 5.7% is thought to be that the powder was filled 3% less than the assumed filling rate, and the shrinkage of the capsule was larger. In addition, the difference between the dimensions predicted from the actual filling rate after HIP and the actual dimensions after HIP is up to 1.2% for the flange and 2.3% for the valve bodies. It is necessary to accurately predict the filling rate at the time of design and manage the actual filling rate according to the design value.

The maximum error difference between both the flange and valve body in the pre-HIP capsule design shape and the actual capsule shape, and the maximum error difference between the post-HIP shape and the respective powder filling rate is approximately 2.5 mm. This is considered to be a small difference due to the size of the product, suggesting that the capsule accuracy before HIP also affects the shape after HIP.

In addition, the valve body had a larger difference between the capsule before HIP and the design shape and the actual shape than the flange. This is thought to be because the number of parts increased and the number of welding points increased due to the complexity of the shape, and the capsule was distorted by the heat of welding. Therefore, it is necessary to restrain the capsule during welding and to correct distortion after welding.

Mechanical Properties of Flange

Method

The flange, after dimension evaluation, was subjected to 2.5 hours of heat treatment at 1050 °C and a tensile test, a Charpy impact test, corrosion test, and micro-structure observation were performed. For the corrosion test, a pitting test using ferric chloride was performed. As to the micro-structure observation, the amount of ferrite and intermetallic compounds and nitrides present were confirmed. Sampling was based on NORSOK standard m-650.

Results

Table 3 shows the results of the tensile test and the Charpy impact test. All test results met the NORSOK standard. Table 4 shows the corrosion test results and micro-structure observation results. Fig. 7 shows a photograph of the microstructure. The amount of corrosive thinning was a maximum of 0.2 g/m^2. The amount of ferrite was 40.9 - 46.3%. Also, it was confirmed the intermetallic compound and nitride were very small amounts.

Table 3. Tensile test and Impact test results

	TENSILE TEST			IMPACT TEST (-46 °C)	
	0.2% proof Stress [MPa]	Ultimate tensile strength [MPa]	Elongation [%]	Absorbed energy [J]	
				Average	Minimum
NORSOK Standard	$\geqq 550$	$\geqq 750$	$\geqq 25$	$\geqq 45$	$\geqq 35$
Flange	615-624	872-876	37	68-171	66-168

Table 4. Corrosion test and Micrographic examination test results

	CORROSION TEST ASTM G48	MICROGRAPHIC EXAMINATION	
	Weight loss [g/m^2]	Ferrite content [%]	Intermetallic phases and nitride precipitates
NORSOK Standard	<4.0	40 - 60	-
Flange	0.1 - 0.2	40.9 - 46.3	Trace

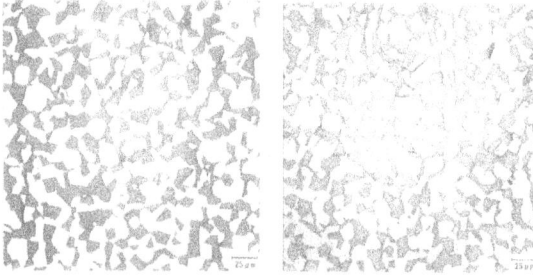

Figure 7. Micrographic of Flange (Left: inner position, Right: surface position)

Mechanical properties of Valve body
Method
After dimension evaluation, the valve body was heat-treated at 1100 °C for 4 hours, and tensile tests, Charpy impact tests, corrosion tests, and micro-structure observations were performed in the same manner as the flanges. Sampling was based on NORSOK standard m-650 with measurements taken at various locations.

Results
Table 5 shows the tensile test and the Charpy impact test results. All parts met the NORSOK standard. Table 6 shows corrosion tests and micro-structure observations. NORSOK standards were met in all respects. Fig. 8 shows microstructure observation photographs of the surface position and inner position. There was no significant difference between the surface and inner portions of the valve body.

Table 5. Tensile test and Impact test results

	TENSILE TEST			IMPACT TEST (-46 °C)	
	0.2% proof Stress [MPa]	Ultimate tensile strength [MPa]	Elongation [%]	Absorbed energy [J]	
				Average	Minimum
NORSOK Standard	$\geq$550	$\geq$750	$\geq$25	$\geq$45	$\geq$35
Valve body	600-633	855-876	37.4-38.4	104-181	96-173

Materials Research Forum LLC
https://doi.org/10.21741/9781644902837-6

Table 6. Corrosion test and Micrographic examination test results

	CORROSION TEST ASTM G48	MICROGRAPHIC EXAMINATION	
	Weight loss [g/m²]	Ferrite content [%]	Intermetallic phases and nitride precipitates
NORSOK Standard	< 4.0	40 - 60	-
Valve body	0.3 - 0.4	42.9 - 44.7	OK

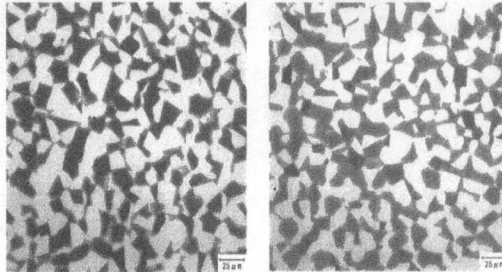

Figure 8. Micrographic of valve body (Left: Inner position, Right: Surface position)

Discussions

Both the flange and the valve body were able to meet the mechanical properties required by NORSOK. As for the heat treatment temperature, in both cases of 1050 °C and 1100 °C, the mechanical characteristics of NORSOK were satisfied. For these reasons, it is considered that even large products such as those using MTC'S Giga-HIP can obtain NORSOK certification.

Conclusions

In the dimensional evaluation, in order to manufacture with NNS, it was found that the powder can be filled according to the filling rate of the design value and the shape management of the capsule can be brought closer to the target shape by filling the powder according to the filling rate of the design value and increasing the production accuracy of the capsule. This indicates that the powder filling rate and the shape management of the capsule are critical.

Since the large flange and valve body were able to meet the mechanical properties of NORSOK, it is thought that even large products such as those using MTC'S Giga-HIP can obtain NORSOK certification.

References

[1] Toyohito Shiokawa, Hiroshi Urakawa, Mitsuo Okuwaki, Yuto Nagamachi "HIP Process of a Valve Body to Near-Net-Shape using Grade 91 Powder" Proceedings of International Conference on HIP, 2017, pp. 58-64.

[2] T. Shiokawa, Y. Yamamoto, S. Hirayama and Y. Nagamachi "Comparison of experimental and FEM simulations of densification during HIP processing of powder into a cylindrical component" Proceedings of International Conference on HIP, 2011, pp. 225-230.

[3] Björn-Olof Bengtsson, "PROGRESS IN THE MATERIAL PROPERTIES OF STAINLESS POWDER FOR NEAR NET SHAPE PARTS. 140605-1012" Proceedings of International Conference on HIP, 2014, 104

Hot Isostatic Pressing - HIP'22   Materials Research Forum LLC
Materials Research Proceedings 38 (2023) 41-47   https://doi.org/10.21741/9781644902837-7

Powder Metallurgy HIP for Naval Nuclear Applications – Trends in Process and Property Development

Colin D. Ridgeway[1,a] *, Terrance Nolan[1,b]

[1]Naval Nuclear Laboratory, Schenectady, NY, USA

[a]colin.ridgeway@unnpp.gov, [b]terrance.nolan@unnpp.gov

Keywords: PM-HIP, Low Alloy Steel, A508 Grade 4N

Abstract. Powder Metallurgy - Hot Isostatic Pressing (PM-HIP) is considered a key technology for component fabrication. By offering near-net shape forming, long lead-time components can be delivered quicker and more efficiently, ultimately supporting on-time construction of nuclear components and structures. To this end, ferrous (A508 Grade 4N) and nickel-base alloys (A600) have been examined in the consolidated PM-HIP condition to assess the mechanical behavior as well as similarity to their wrought/forged counterparts. In this study, various aspects of the PM-HIP process were explored from the powder production to the consolidated material and eventual heat treatment to develop a greater understanding of optimized mechanical properties of PM-HIP material. Trends in processing conditions and various heat treatments were correlated to the performance of each material and related to the wrought counterpart.

Introduction

The United States Navy recently issued its 30-year shipbuilding plan which outlined the number of vessels and platforms to be acquired each year to meet US Navy force-level goals [1,2]. As part of this document, it was identified that in order to meet the future demands of naval nuclear vessel construction (nuclear submarines and aircraft carriers) shipyard output for submarines alone will need to increase by ~250% over the next ~10 years compared to the previous ~10 years. This includes production of the future COLUMBIA-class of ballistic missile submarines which have been identified as the Navy's #1 priority. To meet this demand, the Naval Nuclear Laboratory (NNL) and shipyards have begun to aggressively pursue advanced manufacturing technologies that assure targeted production goals can be achieved and offer the potential to alleviate strain on the current vendor base. PM-HIP has been identified as one such technology and has the potential to displace and/or compliment the current production capacity for naval nuclear components. Apart from enabling manufacturing of critical naval nuclear components, PM-HIP also allows for potential performance/quality improvements and cost/lead time reductions over legacy processes such as casting and forging.

Materials

Materials under consideration for PM-HIP applications include both ferrous and nickel-base alloys, however, the remainder of the discussion will focus on ASTM A508 Grade 4N Class 1 (A508Gr4N), low alloy steel. Two chemistry variants of A508Gr4N, conventional and low residuals, were examined in this study. Powder was sourced from three different vendors who produced the powder using either vacuum inert gas atomization (VIGA) or inert gas atomization (IGA). Powder details are shown in Table 1. It was determined that the IGA process could only achieve conventional chemistries but was still included for powder and mechanical property comparisons.

Table 1 – Powder and HIP details for A508 Grade 4N material analyzed.

Vendor	Target Composition Type[1]	Melt Method	Atomization Gas	Nominal PSD [μm]	Powder Oxygen Content (ppm)	Billet Outgas Temperature	Billet HIP Temperature / Time
A	Low Residuals	VIGA	Nitrogen	0-500	200	70°F, 250°F, 500°F	2065°F – 4 hrs or 2215°F – 4 hrs
	Low Residuals	VIGA	Nitrogen	53-500	140		
B	Low Residuals	VIGA	Nitrogen	53-500	150	70°F, 250°F	2190 – 6 hrs
C	Conventional	IGA + No Cover Gas	Nitrogen	50-150	190	400°F	2250°F – 4 hrs

[1]Low residuals contain low levels of Si, Mn, P and S compared to conventional compositions.

Upon receipt of the material from each vendor, the powder was analyzed via Scanning Electron Microscopy (SEM) and Auger electron Spectroscopy (AES) to understand basic powder characteristics and to characterize the inherent oxide layer on the surface of all particles. Of the two lots of powder from Vendor A, 12 billets (~17"x8"x8") were HIP consolidated (6 from each PSD) with each undergoing a unique processing sequence. This allowed any variation in properties due to particle size distribution (PSD), outgas temperature, and HIP temperature to be uniquely defined. Remaining powder from Vendors B and C was HIP'd under the parameters shown in Table 1 to form billets of 22"x8.5"x8.5" and ~20"x9"x5" respectively.

Following HIP consolidation of all 15 billets, a standard austenization (1575°F – 4 hrs), water quench and temper (1205°F +/-25°F – 10 hrs) was applied to billet sub-sections measuring ~8"x4"x4" with the canister still present. Billet sub-sections were then further sectioned into tensile and Charpy specimens. Tensile specimens were tested at 70°F (per ASTM E8), while Charpy specimens were tested at a range of temperatures (per ASTM E23) to define the ductile to brittle (DTB) transition associated with PM-HIP A508Gr4N.

Results and Discussion

Powder Analysis – Oxide Thickness Correlation

Manufacturing A508Gr4N powder is relatively new to the existing vendor base, and it was desired to characterize the current powder production capabilities. The VIGA and IGA powders obtained from the three vendors was first analyzed under the SEM with images of each powder shown in Fig. 1. Much of the powder produced via the VIGA process was characterized as having spherical or oblong particles with moderate to heavy satellite particles attached to the main particle. Conversely, the IGA powder particles exhibited a more ideal particle geometry consisting of spherical particles with smoother particles surfaces and fewer satellites.

After examining the powder morphologies, each powder was analyzed using AES to characterize the particle and satellite oxide thickness from powder produced at each vendor. AES clarified that there was not a significant variation in the oxide thickness from the VIGA and IGA processes. However, depth profiles indicated that thinner oxides were consistently observed on powder that was spherical in morphology and had fewer satellites, whereas particles with mottled surfaces and increased satellites had comparatively larger oxide thicknesses as shown in Fig. 2. This would suggest that the powder production method has limited bearing on the ability to produce powder with a quality morphology as powder produced using the IGA process possessed the most favorable powder particle morphology and fewest satellites. Variation in powder morphology may be correlated to vendor experience with particular alloys, as Vendor C is known to regularly produce AISI 4340, while Vendors A & B had limited experience producing low alloy steel powder.

When the two PSD as produced from Vendor A were examined, the oxide thickness was found to be consistent between the full cut of powder and larger PSD. Thus, the oxide thickness on all has no correlation to particle size. This corelates with the best practice to screen out the small particles or "fines" due to the increased surface area corresponding with elevated oxygen content that will carry into the HIP consolidated component.

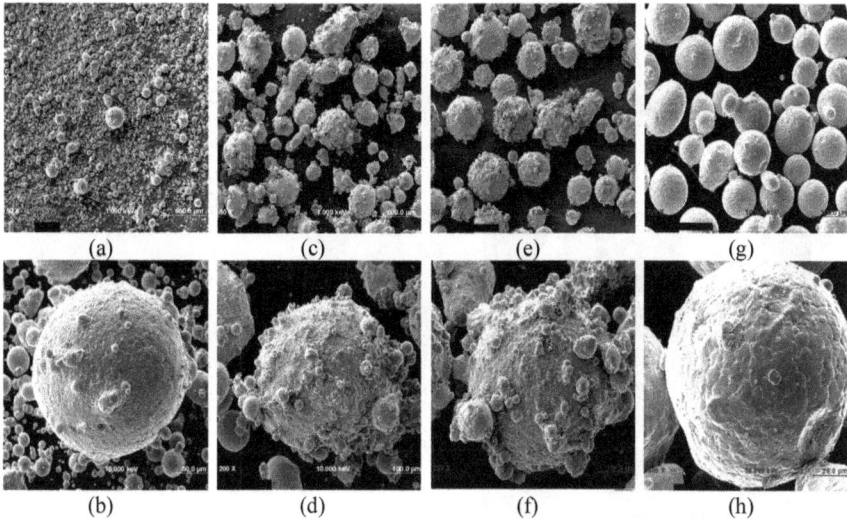

Figure 1. Powder Particles from (a-b) Vendor A – VIGA, full cut powder, (c-d) Vendor A – VIGA, fines removed, (e-f) Vendor B – VIGA, fines removed, and (g-h) Vendor C – IGA, fines removed.

Mechanical Properties of HIP Consolidated Material

Consolidated and heat treated PM-HIP material was mechanically tested to compare PM-HIP properties to the wrought ASTM A508Gr4N equivalent as well as identify the DTB transition curve. Tensile and Charpy specimens were extracted from billet sub-sections in a grid-like pattern with a general extraction location detailed in Fig. 5.

Tensile properties for all billets from all three vendors were found to have room temperature tensile properties that exceeded the minimum requirements for ASTM A508 Grade 4N Class 1 with no single data point falling below the minimum requirements. Tensile properties among the three Vendors had little deviation in strength while maintaining considerable ductility despite the variation in melting/atomization practice and resulting differences in chemistry (conventional vs. low residual). This suggests that the PM-HIP process may be robust enough to recover from slight variations in supplier chemistry, or powder production method, if only tensile properties are required. The robust nature of the tensile properties are highlighted in Fig. 3 which depicts 12 unique processing conditions for Vendor A and shows minimal variation for each condition. Further testing is recommended to confirm this trend, however, the limited variation suggests that hot outgassing or screening out smaller particles is not required to achieve minimum required tensile properties. This has the potential to save costs and reduce production span time.

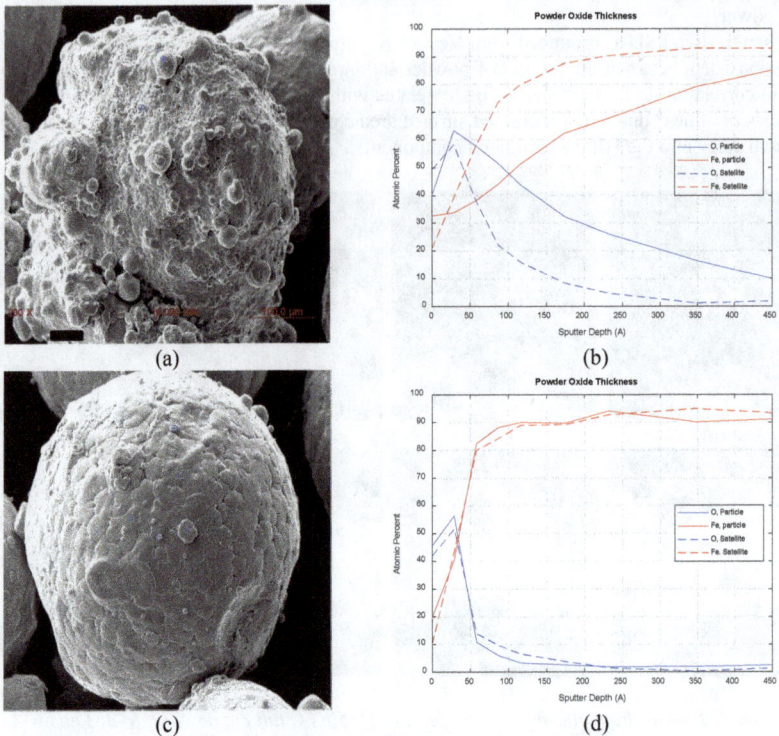

Figure 2. Powder oxide thickness plots for representative (a-b) VIGA produced powder, Vendor A & B and (c-d) IGA Produced powder, Vendor C.

Near Canister Degraded Impact Toughness

The Charpy impact toughness response for each of the billets is shown in Fig. 4. Analysis of the impact data provides two key conclusions (1) there is a significant reduction in properties for the IGA produced powder compared to the VIGA powder and (2) there is a large amount of scatter in the existing data. The reduced impact toughness of the IGA powder is highlighted in Vendor C possessing significantly lower impact toughness which consistently measured more than 50 ft-lbs below the VIGA produced material from Vendor A. The ASTM required average impact toughness for A508Gr4N is 35 ft-lbs at -20°F with no specimens measuring below 30 ft-lbs. Powders from all three vendors exceeded this requirement. Only powder from Vendor A exceeds a notional 100ft-lb target the Electrical Power Research Institute (EPRI) has established for PM-HIP A508 Grade 3 development [3].

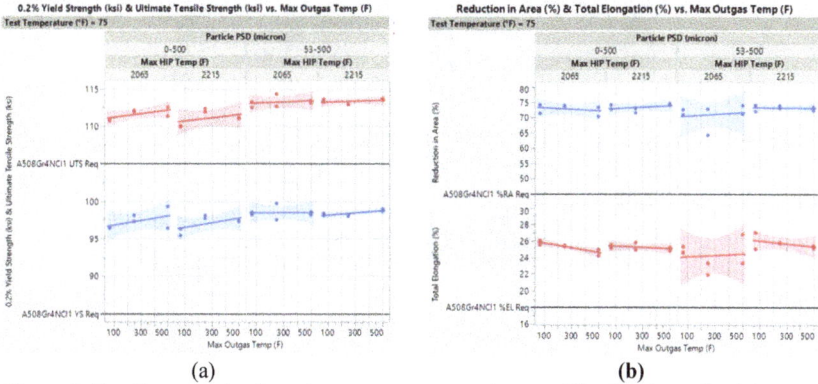

Figure 3. Tensile properties for unique processing conditions of Vendor A powder (a) strength and (b) ductility.

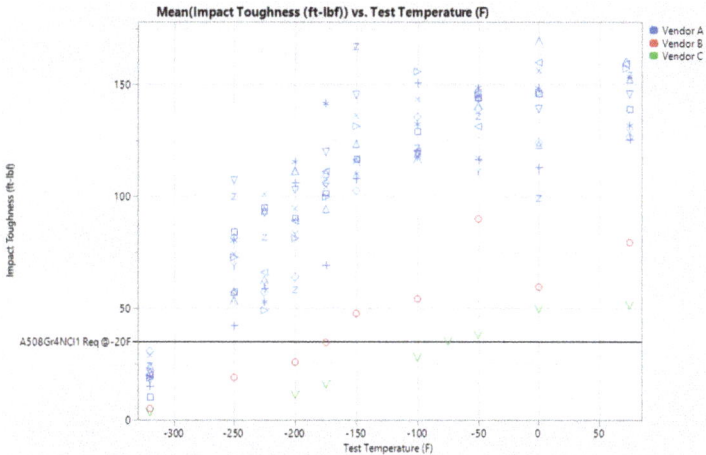

Figure 4. Charpy impact toughness data from each of the three vendors.

The degree of scatter observed in Fig. 4 was not expected and resulted in further analysis of individual data points for each of the billets tested. Conveniently, the grid-like pattern used for Charpy specimen in Fig. 5a allowed for pseudo heat maps to be developed to determine location specific toughness across each billet. Fig. 5b shows up to 100 ft-lbs variation in impact toughness when comparing the specimens located immediately adjacent the original canister (specimens 10-17 in Fig. 5a) compared to the specimens located in the billet interior. By separating the location specific toughness response, the scatter is diminished, and the PM-HIP material exhibits two unique DBT curves which was observed across all vendors and powder types. The specimens extracted from the billet interior were found to have improved impact toughness compared to the near-can region with upper shelf toughness for Vendor A interior billet specimens ranging from 140-175 ft-lbs compared to ~90 ft-lbs for near can regions.

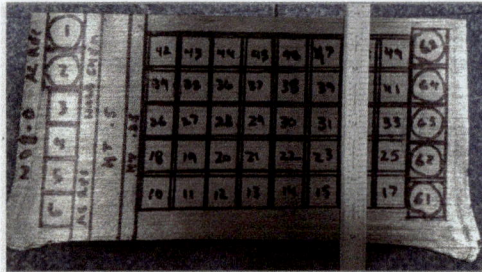

(a)

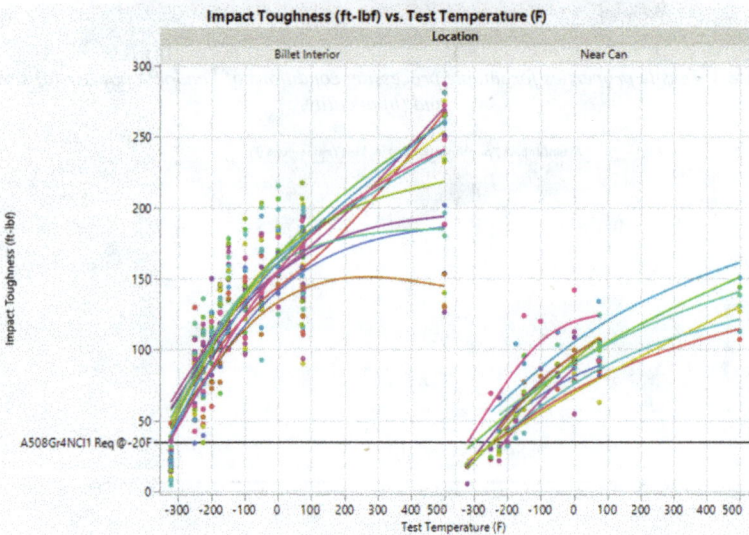

(b)

Figure 5. (a) Charpy and tensile specimen extraction location and (b) DTB Transition curve highlighting the near can response for billets from Vendor A.

The mechanism resulting in the degraded properties in the near-can region are currently not well understood. Fractography of the Charpy specimens fracture surfaces exhibited ductile tearing or microvoid coalescence for both locations and no distinct variation. Kinetic or chemical effects were also ruled out as both μXRF and electron probe microanalysis scans across the canister-billet interface showed no chemical variation. The only variation between the two regions of the PM-HIP billets is the presence of increased oxide inclusions decorating prior particle boundaries in the near-can specimens compared to the billet interior specimens as shown in Fig. 6. The increased number of inclusions creates a network which may result in a preferential cracking pathway and reduced toughness in material up to1-1.5" into the billet interior as measured from the original canister surface. Note that all tensile specimens were extracted from the "near-can" region but still exhibited acceptable properties.

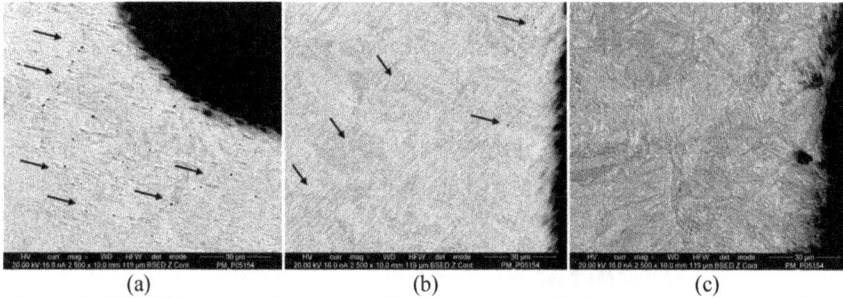

(a)   (b)   (c)

Figure 6. SEM BSE images of near-notch Charpy specimens showing inclusions indicated by black arrows (a) along prior particle boundaries of a near-can specimen (b) scattered in interior billet specimens and (c) forged A508Gr4N microstructure for comparison.

Conclusion

NNL examined consolidated PM-HIP A508Gr4N billets fabricated with both VIGA and IGA powder. The tensile properties of the VIGA and IGA material was found to be equivalent to typical wrought tensile properties despite all tensile specimens being extracted from near canister locations. The Charpy results showed significant variation between Vendor A B and C. All material exceeded the current ASTM A508 Grade 4N Charpy impact requirements, but only Vendor A exceeded a notional 100ft-lb target value at room temperature. General property evaluation suggests that even though quality particles are obtainable via the IGA process, the Charpy impact properties are degraded relative to VIGA powder. Despite different melting and outgas/fill method, all 15 billets examined within this study exhibited a reduction in properties near the canister and was estimated to be ~1-1.5" into the material as measured from the canister. It appears that there may be an increased level of oxides at prior particle boundaries in these regions though results are not conclusive. The mechanism for increased oxide content near the canister is not clear at this time and requires additional investigation.

References

[1] Navy Force Structure and shipbuilding Plans: Background and Issues for Congress. Congressional Research Service, RL32665, April 27, 2022. https://crsreports.congress.gov/product/pdf/RL/RL32665

[2] Report to Congress on the Annual Long-Range Plan for Construction of Naval Vessels for Fiscal Year 2023. Prepared by Office of the Chief of Naval Operations, April 2022. https://media.defense.gov/220152127022/Apr/20/2002980535/-1/-1/0/PB23%20SHIPBUILDING%20PLAN%2018%20APR%202022%20FINAL.PDF

[3] Advanced Nuclear Technology: Code Development of PM-HIP Alloy A508—Progress Report. EPRI, Palo Alto, CA: 2021. 3002018274.

Rapid L-PBF Printing of IN718 Coupled with HIP-Quench: A Faster Approach to Combine Manufacturing and Heat Treatment in a Nickel-Based Alloy

Emilio Bassini[1,2,3,a] *, Giulio Marchese[1,2,3,b], Davide Grattarola[1,c],
Pietro A. Martelli[1,3,d], Sara Biamino[1,2,3,e], Daniele Ugues[1,2,3,f]

[1] Dipartimento di Scienza Applicata e Tecnologia (DISAT) at Politecnico di Torino, Torino, Italy

[2] Consorzio Interuniversitario Nazionale per la Scienza e Tecnologia dei Materiali (INSTM), Firenze, Italy

[3] Integrated Additive Manufacturing@Polito (IAM) at Politecnico di Torino, Torino, Italy

[a]emilio.bassini@polito.it, [b]giulio.marchese@polito.it, [c]davide.grattarola@polito.it,
[d]pietro.martelli@polito.it, [e]sara.biamino@polito.it mail, [f]daniele.ugues@polito.it

Keywords: Laser Powder Bed Fusion (L-PBF), Nickel-Based Alloys, Inconel 718, Heat Treatment, HIP-Quench, Rapid Cooling

Abstract. Additive manufacturing has been attracting more and more interest in recent years. Researchers are constantly involved in enhancing component quality by tailoring the printing parameters, increasing the lead time and decreasing overall productivity. Rapid L-PBF printing is becoming an appealing strategy to make the components manufacturing faster. However, a rapid building strategy will likely cause a higher density of internal flaws that will be healed during the Hot Isostatic Pressing (HIP) cycle. In this work, Inconel 718 was L-PBF printed with two different strategies: The first consisted in creating a dense 1mm shell leaving loosened powders in the core; the latter used two sets of printing parameters for the 1mm shell and the core, respectively. The two strategies lead to 60 and 45% time reduction, respectively. Secondly, full densification and porosity elimination were achieved using a HIP-quench approach, which combines a fast-cooling step at the end of the HIP cycle, eliminating the need for a further annealing treatment. This work shows the results presenting the final microstructures and the retained flaw. Finally, the microstructural degree of recrystallization was also assessed via EBSD analysis.

Introduction

In recent years, powder metallurgy is becoming more and more bound to Hot Isostatic Pressing and Additive Manufacturing (AM). The first one can be used as a Near Net Shape manufacturing technique or as a post-processing step to reduce the presence of defects in components obtained from other processing routes [1]. Additive Manufacturing, on the other hand, with all its sub-categories, has demonstrated a powerful tool for the realization of complex components with a wide variety of materials [2]. HIP is known for generating very fine microstructures with no building textures and enhanced mechanical properties. Nevertheless, it is reported that ductility of HIPped components could be drastically reduced by the presence of Prior Particle Boundaries (PPBs) [3]. Hence, an optimized HIP treatment should always aim to minimize the presence of these harmful species. A promising manufacturing route could foresee the combined use of HIP and AM, aiming to reduce leading times considerably. The AM allows obtaining components with a complex design and shape, which cannot be obtained with traditional manufacturing routes. Nonetheless, the advantages brought by AM become less evident if the printing time is considered. The layer-by-layer strategy can be highly time-consuming. HIP can be introduced to limit this evident drawback of AM. In this paper, IN718 was printed with two strategies for different time reductions. The first one consisted in printing only 1 mm thin walls (shell) during the L-PBF process and leaving the remaining powders unmelted. This strategy reduced the job time by ca. 60%. The geometry of shelled samples was taken from Plessis et al., who successfully printed Ti6Al4V samples[4]. The second strategy consisted of printing only a 1mm shell with optimized parameters, while the remaining inner part was laser printed with a faster scan rate. By doing so, the shell is entirely dense while the part underneath contains a considerably higher content of defects. A perfectly dense shell is fundamental because it seals the inner material from Argon leakages during the HIP allowing an optimal pressure transfer. The printing was 45% faster compared to a traditional job. As a further optimization, the HIP treatment was terminated with a fast-cooling step (HIP-Quench) to obtain extremely small precipitates, settling the best condition for the following ageing treatment [5]. This work aims to understand how the printing strategies impact the final microstructure of the samples, searching for a good compromise.

Materials and methods

Figure 1a shows the commercial gas-atomized IN718 powders produced by EOS used to build the samples; Fig. 1b shows the powder size distribution (PSD) and the relative cumulative curve with an indication of its critical diameters.

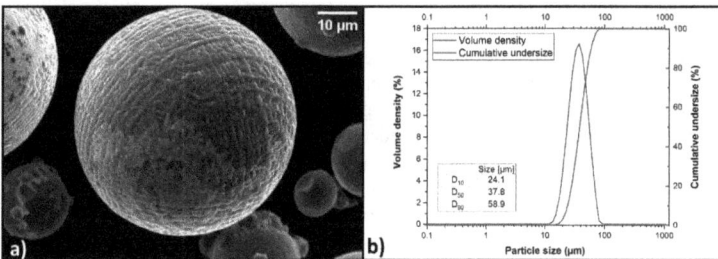

Figure 1. Gas atomized IN718 a) and PSD (blue) and cumulative curve (red) b).

Figure 1a shows spherical particles with an evident dendritic structure. Despite a good circularity, some particles show satellites and splats on their surface. The chemical composition, per UNS N07718, was double-checked via Energy-dispersive X-ray analysis and is reported in Table 1.

Table 1. Chemical Composition of IN718 expressed in wt%; only hydrogen is expressed in ppm

Element	Al	Ti	Cr	Fe	Nb	Mo	O	N	C	H	Ni
Amount [w%]	0.59	1.14	19.3	18	5.99	2.82	0.0181	0.0155	0.0379	1.85 [ppm]	Bal.

The table shows O, N, H and C content assessed using Leco ONH 836 and Leco CS744. These two pieces of equipment rely on melting under inert gas (High purity He) and IR combustion, respectively, to quantify the abovementioned interstitial elements. Powders were loaded into a Mlab Cusing R machine by Concept Laser GmbH equipped with a 100w fibre laser. All the specimens were produced using a constant flow of Argon and without preheating the platform. Specimens were produced using an optimized set of parameters only in the outer part, aiming for faster printing. More specifically, the optimized parameters were used to produce the samples' outer part, hereafter referred to as shell. Parameters come from previous work from our research group [6] and are briefly summarized below. The laser power was set at 95 W; layer thickness was 20 μm; scanning speed was 800 mm/s while hatching distance was 0.05 mm. The samples were produced using a strip width of 5 mm. The scanning pattern was rotated 67° after each layer was completed. The inner part of the samples was produced in two different manners: a group consisted of a hollow shell in which powders were not laser sintered. The latter used two sets of parameters: one for creating the outer part of the sample and one for the inner part. The shell was printed using optimized parameters while the core was printed using a much faster scan rate, i.e. 2400 mm/s leaving all the other parameters unchanged. This sample family will be called "Controlled Porosity" (CP). Fig. 2a shows the 3d CAD schematic of the samples and their most relevant measures.

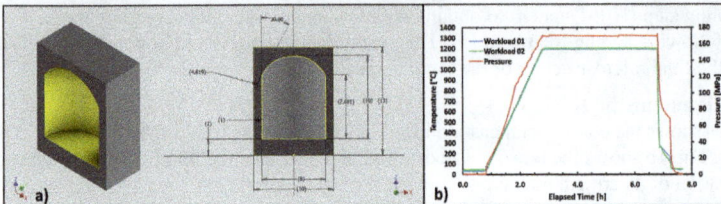

Figure 2. Schematic representation of the samples (measures in mm) a) and the HIP cycle used.

In the schematic above, the grey part is the shell, i.e., the part printed with optimized parameters. Conversely, the part highlighted in yellow contained the powders lasered with a higher scan rate or unsintered, depending on the adopted printing strategy. The HIP cycle to densify the samples was performed in a Quintus QIH 15L molybdenum furnace with a URC module. The cycle was performed according to the work of Rao et al. [7] at 1200 °C for 4 hours but using a higher Ar pressure, i.e., 170 MPa. The URC module was set to cool the samples as fast as possible (270°C/min) to promote the precipitation of fine reinforcing particles and carbides. The cooling rate was measured between 1200 and 400 °C using a thermocouple in close contact with the samples. Fig. 2b shows the plot of the heat treatment; temperature and pressure were raised simultaneously; the heating rate was 10°C/min. After the densification cycle, the samples were sandblasted, and their density measured using a pycnometer Anton Paar Ultrapyc 5000. This initial test confirmed that a satisfactory level of densification was achieved. Nevertheless, a more in-depth procedure to evaluate the porosity level was used. Samples cross sections were embedded in resin and prepared for metallurgical observation. Samples were ground with SiC papers (from 180 to 1200 grit) and polished with diamond pastes (from 6 to 1μm). The correct surface finishing was achieved using colloidal silica. Image analysis was applied to ten light optical microscope pictures to assess the porosity fraction. Microstructures were obtained using electrochemical etching at 3V for 5 seconds. The Electrolyte solution was a 1:3 solution of Nitric and Chloridoid acids. Other reference etching types, such as Kalling #2, failed to reveal the microstructures properly. Optical and scanning electron microscopes (Leica mef4 and Zeiss Evo 15, respectively) were used to characterize the microstructure of the samples correctly and at different scales. Finally, as-polished samples were observed in a Tescan S900G equipped with an EBSD detector to appreciate better the differences in microstructures between the lasered and the HIP-sintered parts.

Results

Table 2 shows the sample density measured immediately after the L-PBF stage and the HIP cycle using the pycnometer.

Table 2. Density levels before and after HIP

Sample type	Initial Density [g/cm^3]	After HIP density [g/cm^3]
Standard L-PBF	8.223 ± 0.002	8.229 ± 0.002
Shelled samples	6.781 ± 0.014	8.221 ± 0.002
CP samples	7.923 ± 0.019	8.225 ± 0.006

IN718 is a material with high weldability, and a density level close to the theoretical one can be achieved when it is printed with optimized parameters. Shelled samples have the lowest density level, while samples with CP remain in the middle. Noteworthy density increases significantly after HIP for the shells and CP samples being practically indistinguishable.

Fig. 3 shows low magnification micrographs of as polished samples to assess better the population of retained porosity of previously mentioned samples.

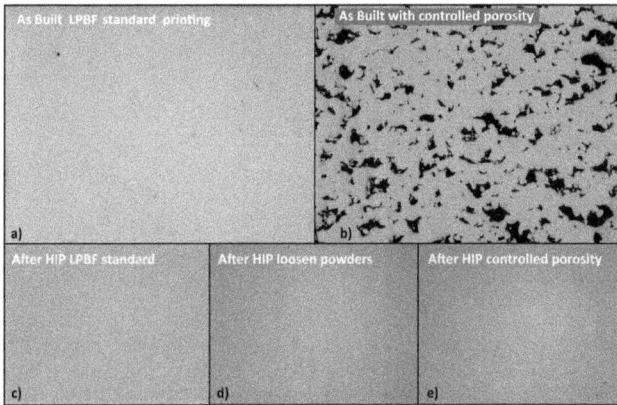

Figure 3. As L-PBF sample printed with optimized parameters a), as-printed CP sample b), traditional LPBF sample after HIP c), shelled sample with loosened powders after HIP d), and CP sample after HIP.

As can be seen, the L-PBF sample printed in the standard condition is practically fully dense; conversely, the CP samples show a large volume of defects such as lack of fusion and pores. Naturally, the shelled sample with loosened powders couldn't be investigated before the HIP stage. The second row of Fig. 3 shows an extremely low density of residual flaws, indicating that densification was completed in accordance with density values previously obtained with the pycnometer. Fig. 4 shows the samples after the HIP treatment with an optical microscope after chemical etching.

Hot Isostatic Pressing - HIP'22
Materials Research Proceedings 38 (2023) 48-54

Materials Research Forum LLC
https://doi.org/10.21741/9781644902837-8

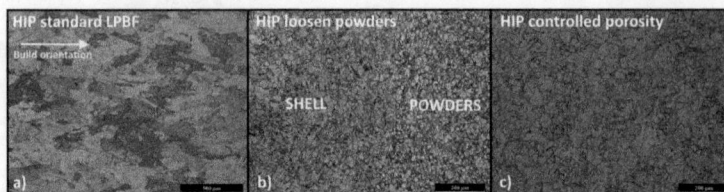

Figure 4. Microstructure of standard L-PBF sample after HIP a) sample with loosened powder after sintering in HIP b) and sample with controlled porosity after HIP c).

The sample obtained with standard L-PBF shows large grains oriented in the building direction; conversely, the shelled sample has a very fine microstructure, resembling that of a sample sintered with a HIP process. Interestingly, the sintering level was good, although the powder's particle size distribution was very narrow, which is usually undesirable for HIPping. Finally, the HIPped sample with controlled porosity shows a hybrid microstructure. Grains are slightly larger than the previous ones but considerably smaller than those shown in Fig. 4a. Fig. 5, taken with backscattered electrons, can be used to explain these differences.

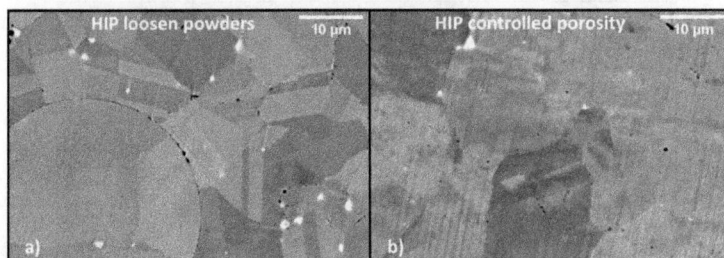

Figure 5. Backscattered electrons images of shelled sample a) and CP samples after HIP.

Figure 5a shows the shelled sample after HIP and PPBs can be observed. These particles hindered the grains from passing across the particles. PPBs are mixed oxides and carbides formed on the particle surface and appear as small black points in the micrographs. Conversely, the white spots are Nb carbides, clearly visible in both sample typologies. PPBs presence and the complete absence of laser exposure during the printing stage allowed to maintain the grain extremely small. On the other hand, Fig. 5b shows oxides and carbides, which originally decorated the spherical powders, dispersed throughout the microstructure. Grains could cross the particles due to a less effective barrier effect of the PPBs. This phenomenon was probably caused by the laser interaction, which caused the partial melting of the powders. PPBs were not dissolved due to their extreme stability but were re-arranged more diffusely. Fig. 6 shows PPBs at higher magnification, and the resulting EDS performed both on the matrix and oxy-carbides.

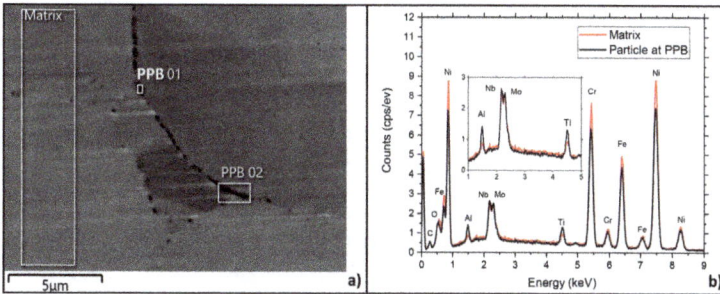

Figure 6. PPBs detail from SEM micrography a) and the resulting EDS spectrum (black curve) compared to matrix (red curve) b).

According to the EDS results, the particles that nucleates at powder boundaries are richer in Al and Ti, both elements which are known to segregate towards the powder surface and are prone to react with Oxygen or Carbon. As a result, highly stable oxy-carbides with complex stoichiometry are formed and remain unaltered even at HIPping temperatures, as mentioned by [3]. Fig. 7 was taken with EBSD at the interface between the L-PBF shell and the part immediately below. Again, the differences between the sample with sintered loosen powders Fig. 7a and the sample with controlled porosity Fig. 7b are rather evident.

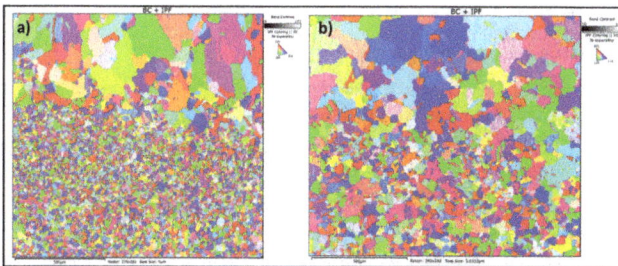

Figure 7. EBSD image of shelled sample a) and CP sample b)

The EBSD images confirm what was previously suggested by other images. Pure L-PBF parts, i.e., the shells, are free to coarsen during the HIP treatment. Despite this ability, a certain degree of texture along the building direction is still appreciable, thanks to the IPF maps. This texture disappears underneath the shell in the sample with loosened powders. Conversely, the CP sample shows a slightly higher residual texture below the shell because of the partial melting and solidification of the powders after the laser passage.

Conclusions

According to this preliminary study, it is possible to draw the following conclusions regarding the possibility of building shelled samples or samples with controlled porosity via L-PBF. First, the density levels obtained in both the sample types are similar and indistinguishable from standard HIPped L-PBF samples. Nevertheless, shelled samples require to predict the shrinkage caused by the sintering. On the other hand, samples with controlled porosity do not show any significant shrinkage. Secondly, although shelled samples make the building rate 60% faster, they may suffer from reduced ductility due to the high fraction of PPBs. Conversely, CP samples lead to a 45%

faster process, but the resulting microstructure looks more promising due to the lower PPBs fraction. Based on this microstructural assessment, the controlled porosity strategy seems to be a good compromise between the reduction of time and the presence of potentially harmful species.

References

[1] D. Herzog, K. Bartsch, B. Bossen, Productivity optimization of laser powder bed fusion by hot isostatic pressing, Additive Manufacturing. 36 (2020) 101494. https://doi.org/10.1016/j.addma.2020.101494

[2] E. Bassini, A. Sivo, P.A. Martelli, E. Rajczak, G. Marchese, F. Calignano, S. Biamino, D. Ugues, Effects of the solution and first aging treatment applied to as-built and post-HIP CM247 produced via laser powder bed fusion (LPBF), Journal of Alloys and Compounds. 905 (2022). https://doi.org/10.1016/j.jallcom.2022.164213

[3] L. Chang, W. Sun, Y. Cui, R. Yang, Preparation of hot-isostatic-pressed powder metallurgy superalloy Inconel 718 free of prior particle boundaries, Materials Science and Engineering A. 682 (2017) 341–344. https://doi.org/10.1016/j.msea.2016.11.031

[4] A. du Plessis, B. Yelamanchi, C. Fischer, J. Miller, C. Beamer, K. Rogers, P. Cortes, J. Els, E. MacDonald, Productivity enhancement of laser powder bed fusion using compensated shelled geometries and hot isostatic pressing, Advances in Industrial and Manufacturing Engineering. 2 (2021) 100031. https://doi.org/10.1016/j.aime.2021.100031

[5] W. Wang, Z. Chen, W. Lu, F. Meng, T. Zhao, Heat treatment for selective laser melting of Inconel 718 alloy with simultaneously enhanced tensile strength and fatigue properties, Journal of Alloys and Compounds. 913 (2022) 165171. https://doi.org/10.1016/j.jallcom.2022.165171

[6] R. Barros, F.J.G. Silva, R.M. Gouveia, A. Saboori, G. Marchese, S. Biamino, A. Salmi, E. Atzeni, Laser Powder Bed Fusion of Inconel 718: Residual Stress Analysis Before and After Heat Treatment, Metals. (2019). https://doi.org/10.3390/met9121290

[7] G.A. Rao, M. Srinivas, D.S. Sarma, Effect of oxygen content of powder on microstructure and mechanical properties of hot isostatically pressed superalloy Inconel 718, Materials Science and Engineering A. 435–436 (2006) 84–99. https://doi.org/10.1016/j.msea.2006.07.053

Materials Research Forum LLC
https://doi.org/10.21741/9781644902837-9

Use of High-Pressure Heat Treatment (HPHT™) for L-PBF F357

Chad Beamer[1,a*], Andrew Wessman[2,b] and Donald Godfrey[3,c]

[1] Quintus Technologies LLC, Lewis Center, OH

[2] The University of Arizona, Department of Materials Science and Engineering, Tucson, AZ

[3] SLM Solutions Americas, Wixom, MI

[a]chad.beamer@quintusteam.com, [b]wessman@arizona.edu,
[c]donald.godfrey@slm-solutions.com

Keywords: High Strength Aluminum Alloys, F357, Hot Isostatic Pressing, HIP, High Pressure Heat Treatment, HPHT, Uniform Rapid Quenching, URQ, Hydrogen Blistering, Distortion, Additive Manufacturing

Abstract. Recent advancements in hot isostatic pressing (HIP) equipment now offer the ability to integrate HIP and heat treatment in the HIP furnace with the aid of controllable high-speed cooling and in-HIP quenching. This approach not only offers improvement in productivity but provides a path to prevent anomalies during heat treatment including thermally induced porosity (TIP) and part quench cracking or distortion. This manuscript will cover the approach of High Pressure Heat Treatment (HPHTTM) applied to SLM Solutions laser powder bed fusion (L-PBF) printed high strength aluminum alloy F357. Microstructure, tensile properties, and part distortion are evaluated. The results capture a post process method making it possible to prevent hydrogen blistering, mitigate defects present in the L-PBF material, offer strength properties exceeding that of MMPDS cast properties, and minimize geometric distortion of complex part geometries.

Introduction

High strength aluminum alloys are of great value to the aerospace industry and their market share reflects this with a revenue share of ~31% reported in 2019 [1]. The demand is due to the advantages offered such as excellent strength-to-weight and stiffness-to-weight ratios, good machinability and formability, corrosion resistance, and low-cost compared to more exotic materials. As a result, these alloys are of wide use for non-critical and structural applications offering improvements in fuel efficiency and reduction in weight of the aircraft [2].

F357, also commonly referred to as AlSi7Mg, is a high-strength aluminum alloy supporting this demand. Traditionally it's a cast precipitation hardenable Al-Si-Mg alloy that has been extensively studied. Its increase in Mg over AlSi10Mg offers improved balance in mechanical properties even at elevated temperatures. However recent development interest has been placed for its use in additive manufacturing (AM) techniques including L-PBF.

AM technology is expanding the design window providing the possibility to manufacture more innovative shapes and complex geometries not possible by conventional fabrication techniques. New sets of design freedoms are enabling lower production cost and part failure risk, improved performance including lightweighting, and lower material usage as part complexity increases. This is creating additional interest and use for high-strength aluminum alloys including F357 in the aerospace industry. In fact, AM as it has been marked with significant growth rates in recent decades, doubling the global AM market size every three years through 2023. And it's the aerospace industry that has been leading the charge in AM development for metallic non-critical and mission-critical components including aluminum alloys [3].

Despite these advantages and opportunities that exist there has been a slow adoption of AM high-strength aluminum alloys in the aerospace industry. This is largely due to the print defects inherent to the L-PBF process impacting part performance and reliability and lack of quality standards for AM components [3, 4].

Lack-of-fusion and keyhole defects are one of the most common print related anomalies present in AM processes. These defects can negatively impact microstructure and mechanical properties limiting structural integrity and use [4]. A method commonly used to minimize or eliminate such defects is hot isostatic pressing (HIP). However it has been reported in literature that hydrogen porosity and blistering is a common issue in aluminum castings and L-PBF components [5, 6, 7]. The hydrogen porosity and blistering present in castings can be managed with appropriate melt strategies. Due to the rapid solidification of L-PBF such strategies can't be applied, and it has been reported that the hydrogen supersaturated state in the solidified alloy is further enhanced. So despite the HIP process providing a fully dense structure, the need to perform conventional post-processing thermal treatments, such as the T6 solution anneal treatment at elevated temperatures, hydrogen pores may nucleate and grow [5]. These anomalies deteriorate the high density achieved by the HIP process and can negatively impact the microstructure and mechanical properties of the material.

An additional challenge associated with the heat treatment of AM high-strength aluminum alloys is often the requirement to quench from the solution anneal temperature in order to achieve the full age-hardening potential of the alloy. Ideally, the quench is faster than the critical cooling rate, but slow enough to avoid part distortion and even cracking. Often in practice this is not the case. Historically a water quench strategy has been applied to F357 castings resulting in optimum mechanical properties despite the propensity to distortion often requiring rework or scrap [8]. AM F357 is faced with the same challenge if there is a desire to meet or exceed mechanical property requirements of cast variants.

Due to the challenges highlighted above there has been a lack of consistency with post process heat treatment for high-strength aluminum alloys including F357. The benefits offered by HIP can be eliminated by hydrogen blistering. Also, water quench applied from the T6 heat treatment can lead to distortion. Performing a stress relief only heat treatment or direct age offers an approach to avoid both hydrogen blistering and distortion but doesn't take advantage of the age-hardening potential of the alloy, thus mitigating lightweighting potential.

Therefore, there exists a need in the industry to define a standardized and robust post-processing approach. Tocci et. al. [9] investigated the use of a novel heat treat strategy applied to AM and AlSi10Mg to achieve densification and hardening effects typical of T6 heat treatment in one step. The following work captures a case study carried out by University of Arizona, SLM Solutions, and Quintus Technologies expanding on this strategy, evaluating the application of HPHT in a Quintus Technologies' Uniform Rapid Quenching (URQ®) furnace for L-PBF F357. Density, microstructure, mechanical properties, and distortion of this novel approach are evaluated. Results capture a post processing method offering excellent tensile properties while avoiding thermally induced porosity and distortion.

High Pressure Heat Treatment – An Introduction

Quintus Technologies is a leader in high pressure technology and has introduced the next generation of HIP systems providing capabilities beyond conventional densification. Decade's worth of advancements in equipment design, system functionality, and control now offers a path to perform HIP and heat treatment in a combined cycle, referred to as HPHT™. This novel methodology provides a more sustainable processing route with improved productivity and energy efficiency. It also offers opportunities to further optimize microstructures for improvement in material properties coupled with ease of manufacturability.

Typical thermal processes performed on AM materials include HIP, homogenization, solution heat treatment, rapid air cooling or quenching, aging, and tempering. With Quintus' high strength wire-wound pressure vessels it is possible to perform operations all in the HIP. A thinner vessel forging is enabled by high strength wire winding. The vessel also incorporates cooling channels adjacent to the forging and wire-winding package for a more effective heat extraction from the system. The novel design continues internally to the pressure vessel with the ability to perform forced convection cooling, or the mixing of hot and cold process gases, with the aid of fans or ejectors. Such a system offers the ability to perform controllable Uniform Rapid Cooling (URC®) or URQ®. URC increases the natural cooling speed by up to a factor of 10-15, while stirring the gas so the cooling is more uniform in the hot zone. URQ is very fast cooling (>2000°C/min gas rate with cycle load down to 600°C), allowing quenching of most materials in the same or higher rate as commercial oil quench equipment. Quintus delivers both techniques with the option of controlled cooling. An additional benefit of the highly pressurized gas involved in the cooling segment produces an entire production load that will receive more or less the same cooling rates. Originally introduced for productivity improvement, these technologies now offer the cooling rates, accuracy, and uniformity to perform many of these post processing steps all in a HIP vessel.

There are many reported benefits by applying the combined HPHT route such as reduced number of process steps, reduced cycle time and lead time, and improved process and quality control. Other advantages include spending less time at elevated temperatures helping to preserve the fine grain AM microstructure by minimizing grain growth. It has also been shown that by carrying out all elevated temperature process steps under pressure thermally induced porosity can be prevented or mitigated. Finally, manufacturability can be improved through HPHT as this approach reduces the cooling or quench severity during cooling segments which can often lead to part distortion or cracking. The lower temperature differential between the components and gas temperature compared to conventional approaches and application of high pressure both aid in lowering thermal gradients and stresses with a component.

Materials and Methods

F357 cylindrical samples were printed by SLM Solutions. The machine used for this project was an SLM 280 2.0 with a build envelop of 280 mm X 280 mm X 365 mm (11.0 in X 11.0 in x 14.4in) using standard print parameters including 30-micron layer thickness. The machine was equipped with a twin laser system of 400 watts each. In this study, as-built, stress relief only, and five HIP variants were investigated, see Table 1 for a summary of stress relief and HIP variant conditions. All heat treated samples were stress-relieved at 285°C (545°F) for 2 hours and cooled at a rate equal to air cooling or faster. For the baseline variant, HIP was performed at 515°C (959°F) at 100 MPa (14.5 ksi) for 3 hours with a slow cool (10°C/min(18°F/min)) followed by a standard T6. The standard T6 consisted of a solution heat treatment at 530°C (986°F) for 6 hours in air followed by a water quench and an age cycle at 160°C (320°F) for 6 hours. For variants 4.1-4.4 all HIP cycles represented a HPHT strategy by combining the HIP, solution heat treatment, and quench all in the HIP vessel under high pressure with the use of the URQ furnace. Temperature and pressure were modified for these variants targeting 500°C (932°F)/530°C (986°F) and 100 MPa (14.5 ksi)/170 MPa (24.7 ksi). The lower temperature HPHT variants performed at 500°C (932°F) was evaluated to assess the impact HIPing at lower temperatures had on maintaining a finer AM microstructure, whereas the high-temperature variants performed at 530°C (986°F) was evaluated to assess the impact on achieving an optimum supersaturated solid solution state of hardeners. The high-pressure variants performed at 170 MPa (24.7 ksi) were assessed to evaluate the influence increased cooling power offered with the higher pressure and density gas medium.

Table 1. HIP variants, all samples were stress-relieved at 285°C (545°F) for 2 hours and cooled at a rate equal to air cooling or faster.

Variant	HIP Parameters	T6/Age Specifics
Baseline	515°C/100MPa/3 Hrs + Slow Cool	Standard T6, solution heat treat in air furnace, 530°C/6hrs + Age cycle
4.1	530°C/100MPa/3 Hrs + URQ	Age cycle only, 160°C/6hrs
4.2	500°C/100MPa/3 Hrs + URQ	Age cycle only, 160°C/6hrs
4.3	530°C/170MPa/3 Hrs + URQ	Age cycle only, 160°C/6hrs
4.4	500°C/170MPa/3 Hrs + URQ	Age cycle only, 160°C/6hrs

Results

The microstructure for each variant was evaluated. In general variants 4.1-4.4 showed similar microstructures with respect to shape and size distribution of the eutectic silicon. However, variants 4.1-4.4 showed a finer distribution of eutectic silicon relative to the Baseline variant. Micro photos are provided in Fig. 1 for the Baseline and 4.3 variant.

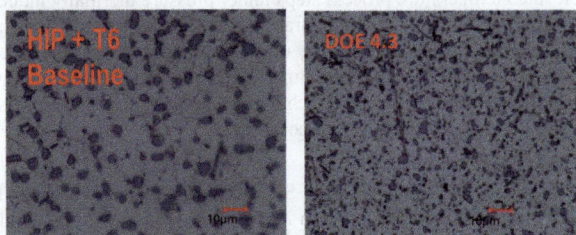

Figure 1. Micro photos of Baseline and 4.3 variants.

ASTM E8 standard tensile coupons were machined and room temperature tensile properties were evaluated for the as-built, stress relief only, as well as the five HIP variants. Results for the 0.2% yield strength (YS) and ultimate tensile strength (UTS) were compared to Metallic Materials Properties Development and Standardization (MMPDS) Cast A357/F357 property data [10] as a benchmark with a summary provided in Fig. 2. Similar trends are observed for YS and UTS. HIP variants 4.1-4.4 display a significant advantage over the as-built and stress relief only conditions, with Baseline outperforming all variants. HIP variants 4.1-4.4 exceeded the benchmark value, with HIP variants 4.1 and 4.3 showing improved strength over the other HPHT variants due to the higher temperature soak segment in the HIP vessel. Elongation for Variants 4.1-4.4 was greater than 15%, equivalent to or better than the Baseline variant. Overall HIP Variants 4.1-4.4 specimen showed no evidence of blistering and negligible distortion relative to water quench samples.

Distortion for HIP Variant 4.1 HPHT cycle was evaluated on a L-PBF F357 trial part shown to be distortion sensitive to the standard T6 water quench. Normalized distortion results are summarized in Fig. 3 for two locations prone to distortion. Overall, the stress-relieved and HIP conditions show little to no increase in dimensional deviation. The as-HT material, comparable to the Baseline variant in this study that utilized a water quench after the solution treatment, shows a significant increase in dimensional deviation. However, the HIP Variant 4.1 mitigated distortion with results similar to stress relieved or As-HIP conditions.

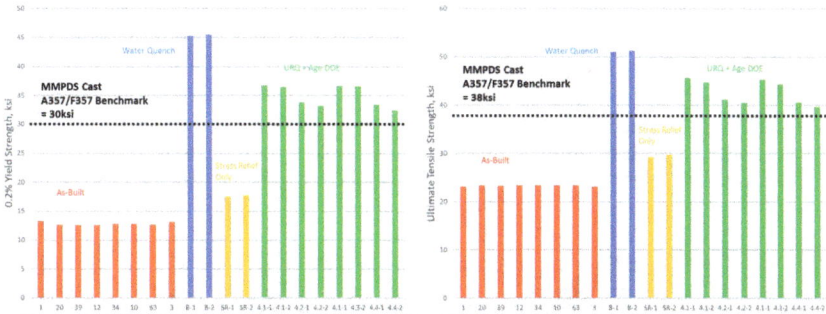

Figure 2. 0.2% YS and UTS results for "As-Built" (sample ID 1, 20, 39, 12, 34, 10, 63, 3), "Water Quench" (sample ID B-1, B-2), "Stress Relief Only" (sample ID SR-1, SR-2), and "URQ + Age DoE" consisting of the URQ HIP variant results compared to MMPDS Cast A357/F357.

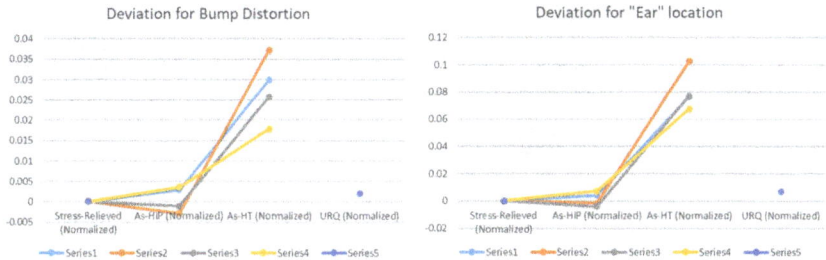

Figure 3. Distortion data at two locations for a distortion-sensitive L-PBF F357 trial part for multiple post-processing conditions.

Conclusions and Future Work

The recent advancements in modern HIP design and equipment are now providing the opportunity to perform HPHT with aid of high-pressure rapid cooling and quenching. The added variable of pressure creates novel approaches to combine HIP and heat treat all in the HIP vessel. This method not only eliminates a process step(s) but also has the potential to remove defects and optimize the microstructure for improved mechanical properties. In this work HPHT was applied to high-strength precipitation-hardenable aluminum alloy F357 produced by L-PBF. These results capture a HPHT processing route applied to L-PBF F357 offering material with no observable defects, strength values exceeding MMPDS cast F357/A357 benchmark properties with excellent ductility, and negligible distortion of a quench sensitive part.

Future work will include post-process optimization studies with a focus on the HIP/solution segment, material characterization of the silicon particle size and size distribution, and ultrasonic low cycle fatigue and high cycle fatigue screening.

Acknowledgements

Collaborations are instrumental in accelerating the knowledge and development of AM technologies and materials. Therefore, we would like to acknowledge Brian Baughman, Manufacturing Chief Engineer at Honeywell Aerospace, for the support and collaboration.

References

[1] High Strength Aluminum Alloys Market Analysis, 2022, Grand View Research, Inc, USA

[2] Daniel F.O. Braga, S.M.O. Tavares, Lucas F.M. da Silva, P.M.G.P. Moreira, Paulo M.S.T. de Castro, Advanced design for lightweight structures: Review and prospects, Progress in Aerospace Sciences, Volume 69, 2014, Pages 29-39, ISSN 0376-0421, https://doi.org/10.1016/j.paerosci.2014.03.003

[3] Sadettin C. Altıparmak, Victoria A. Yardley, Zhusheng Shi, Jianguo Lin, Challenges in additive manufacturing of high-strength aluminium alloys and current developments in hybrid additive manufacturing, International Journal of Lightweight Materials and Manufacture, Volume 4, Issue 2, 2021, Pages 246-261, ISSN 2588-8404, https://doi.org/10.1016/j.ijlmm.2020.12.004

[4] Michael Gorelik, Additive manufacturing in the context of structural integrity, International Journal of Fatigue, Volume 94, Part 2, 2017, Pages 168-177, ISSN 0142-1123, https://doi.org/10.1016/j.ijfatigue.2016.07.005

[5] Van C. Pierre, Krishna M. Gokula, Bigot P. Lore et al., Reducing Hydrogen ores and Blisters by Novel Strategies and Tailored Heat Treatments for Laser Power Bed Fusion AlSi7Mg0.6, Euro PM2019, AM Properties, Al Alloys, https://www.researchgate.net/publication/336649885

[6] Huan Zhao, Poulami Chakraborty, Dirk Ponge, et al., Hydrogen trapping and embrittlement in high-strength Al alloys. Nature 602, 437–441 (2022). https://doi.org/10.1038/s41586-021-04343-z

[7] Daniel Diehl, Formation of hydrogen blisters during the solution treatment of aluminum alloys, Tecnol Metal Mater Min. 2021;18:e2374, https://doi.org/10.4322/2176-1523.20212374

[8] D.S. MacKenzie, N. Bogh, and T. Croucher, Quenching of Aluminum Alloys, Heat Treating of Nonferrous Alloys, Volume 4E, ASM Handbook, ASM International, 2016, p 148–178.

[9] Tocci, M., Pola, A., Gelfi, M. et al. Effect of a New High-Pressure Heat Treatment on Additively Manufactured AlSi10Mg Alloy. Metall Mater Trans A 51, 4799–4811 (2020), https://doi.org/10.1007/s11661-020-05905-y

[10] Battelle Memorial Institute, "Metallic Materials Properties Development and Standardization (MMPDS-16)". Battelle Memorial Institute, 2021.

Hot Isostatic Pressing - HIP'22
Materials Research Proceedings 38 (2023) 61-66

Materials Research Forum LLC
https://doi.org/10.21741/9781644902837-10

Use of HIP Process in Post-Processing of Components Manufactured by SLM Technology from Magnetically Soft FeSi6.5 Powder

Dariusz Kołacz[1,a*], Adrian Radoń[1,b], Karol Krukowski[1,c], Joanna Kulasa[1,d], Aleksandra Kolano-Burian[1,e]

[1]Łukasiewicz Research Network - Institute of Non-Ferrous Metals, Sowińskiego 5, 44-100 Gliwice, Poland

[a]dariusz.kolacz@imn.lukasiewicz.gov.pl, [b]adrian.radon@imn.lukasiewicz.gov.pl, [c]karol.krukowski@imn.lukasiewicz.gov.pl, [d]joanna.kulasa@imn.lukasiewicz.gov.pl, [e]aleksandra.kolano-burian@lukasiewicz.gov.pl

Keywords: HIP, Powder FeSi6.5, Magnetically Soft Material, Process of Post-Processing

Abstract. This paper presents the influence of the Hot Isostatic Pressing (HIP) process on the structure and properties of the components printed by Selective Laser Melting (SLM) technology. The samples were manufactured from magnetically soft FeSi6.5 alloy powder, using different SLM printing parameters. Semi-finished products were densified in the HIP process by varying its duration and isostatic pressure value. Density, hardness, and microstructure were tested on the printed parts both before and after HIP densification. The application of the hot isostatic pressure densification process indicates a change in its final properties concerning the feedstock components. The influence of variable SLM printing parameters of semi-finished parts on the production quality and properties after the HIP process was also observed.

Introduction

Constant technological development generates a demand for new components dedicated to electrical devices. Their efficiency, cost and effectiveness largely depend on the materials used in their construction. Considering, among others, electric motors, transformers, generators, or induction filters, the improvement of highly efficient magnetic materials is of particular importance [1-3]. One of them, which is part of the construction of modern electrical devices and can visibly improve their efficiency, is the Fe-Si alloy, which is classified as soft magnetic material. Due to high electrical resistivity, very low magnetostriction and low magnetocrystalline anisotropy, it has a high potential of being applied in new electrical devices [4]. Unfortunately, due to its fragility, it is challenging to manufacture finished parts with complex shapes by conventional methods. Currently, the progress in Additive Manufacturing technologies allows for the production of ready-made elements using various additive technologies without the need to conduct complicated and energy-consuming processing [2,4].

In the SLM (Selective Laser Melting) process, stresses arise in the printed element (e.g., thermal shock), which may cause their disintegration (e.g., cracking and crumbling). For this purpose, heat treatment is often used to remove stresses in the material and, consequently, improve the magnetic properties [4]. An important factor influencing the mentioned properties is the appropriate structure of the material. For this purpose, additional HIP treatment was performed on the elements printed with the SLM technique. The research presented in the article is a preliminary assessment of the possibility of producing soft magnetic materials FeSi6.5 using a combination of additive technologies and hot isostatic pressing. Materials intended for research in this work were manufactured using the SLM technology and various configurations of printing parameters [5]. Furthermore, the article presents an assessment of the impact of parameters (pressure, time) of post-process treatment by HIP on printed elements made of FeSi6.5 alloy made in various strategies (laser power, scanning speed) of SLM printing.

Research methodology and material

The samples were prepared using SLM 125 device from SLM Solutions. The printed samples were post-processing processed using HIP AIP8-30H device from American Isostatic Press, Inc.

The powder morphology was examined on a JEOL JXA-8230 microscope. The density was determined by the Archimedes method, using ethyl alcohol with a chemical purity of 99.98%. HV hardness was determined using a load of 1 kgf for 15 seconds. Microstructural studies were performed on the Olympus GX71F metallographic microscope and the Zeiss Evo MA10 scanning electron microscope. Phase testing was performed on the Seifert-FPM XRD7 device using the characteristic Cu Kα X-rays and the Ni filter as well as Seifert and Match software and ICDD PDF-4 + catalogue data from 2021.

The samples were prepared using SLM technology from FeSi6.5 alloy powder with a particle size below 100 μm (Fig. 1a). The powder was purchased from PMCtec GmbH (Precision Materials Chemicals and Technology). Examination of its morphology shows the presence of a mixture of globular and spherical particles. Before its use, the alloy powder was subjected to the process of removing moisture by isothermal annealing at the temperature of 120°C for 3 hours and to sieving to obtain the 20 - 63 μm fraction.

(a)   (b)

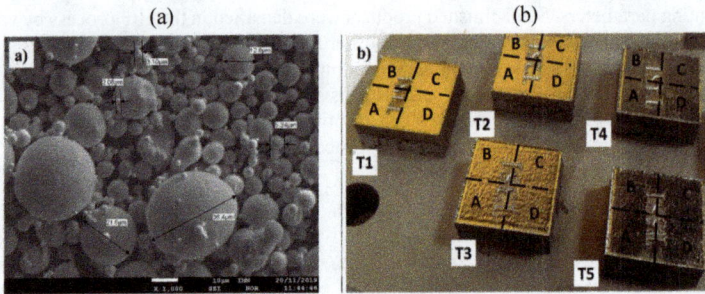

Figure 1. (a) SEM morphology image of the FeSi6.5 powder with particle size below 100 μm, (b) Photo of samples immediately after the printing process with visible support material made of 316L steel with marking of the sample cutting pattern

The samples were produced using five printing strategies, marking them sequentially T1, T2, T3, T4, and T5. The applied printing parameters and their values, along with the assigned markings, are summarized in Table 1. To avoid the printed elements' oxidation, the printing process was carried out in an argon atmosphere. The printing platform was heated to the temperature of 200°C to minimize the phenomena of thermal shocks and their influence on the initiation of possible cracks or delamination in the material. The produced printed elements were in the form of cuboidal samples with dimensions of 20 x 20 x 5 mm (Fig. 1b).

Table 1. Printing parameters assigned to the strategy

Strategy		T1	T2	T3	T4	T5
Laser power	[W]	180	200	220	200	220
Scanning speed	[mm/s]	600	667	733	571	629
Hatch distance	[mm]	0.12				
Layer thickness	[mm]	0.03				
Layer cooling time	[s]	15				

Each of the printed elements was cut into four parts and marked as A, B, C, and D (Fig. 1b). Samples with the A marking were considered as reference material samples, while the remaining samples were intended for the hot isostatic pressing process. HV1 density and microhardness were measured on all samples, both before and after the HIP process. The marking of the tested samples along with the parameters used in the hot isostatic pressing process are summarized in Table 2.

Table 2. Results of density and hardness measurements together with applied HIP process parameters

Sample		T1			T2			T3			T4			T5		
		1B	1C	1D	2B	2C	2D	3B	3C	3D	4B	4C	4D	5B	5C	5D
Before HIP	Density [g/cm³]	7.46 ± 0.02			7.48 ± 0.01			7.46 ± 0.05			7.47 ± 0.03			7.47 ± 0.03		
	Hardness HV1	428 ± 19			424 ± 11			411 ± 15			404 ± 14			431 ± 16		
HIP parameters	Temperature [°C]	1200														
	Pressure [MPa]	100	200		100	200		100	200		100	200		100	200	
	Time [s]	240		480	240		480	240		480	240		480	240		480
After HIP	Density [g/cm³]	7.46	7.47	7.47	7.47	7.46	7.46	7.47	7.46	7.47	7.48	7.46	7.46	7.48	7.46	7.46
	Hardness HV1	399	400	401	409	398	400	402	392	401	392	400	393	398	399	404

The course of the Hot Isostatic Pressing process corresponded to the pre-set parameters, which is illustrated by the temperature and isostatic pressure curves (Fig. 2). In all cases, the pressure was applied simultaneously with increasing temperature. The hot isostatic pressing process was carried out at different values of isostatic pressure (100 MPa, 200 MPa) and the duration of the process (240 and 480 minutes).

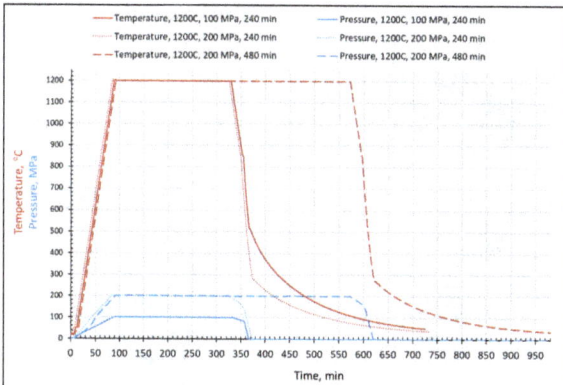

Figure 2. Course of the Hot Isostatic Pressing process

Due to the extensive microstructural research, the article presents the results for the maximum parameters of the HIP process, namely for the parameters: 1200°C, 200 MPa, and 480 min.

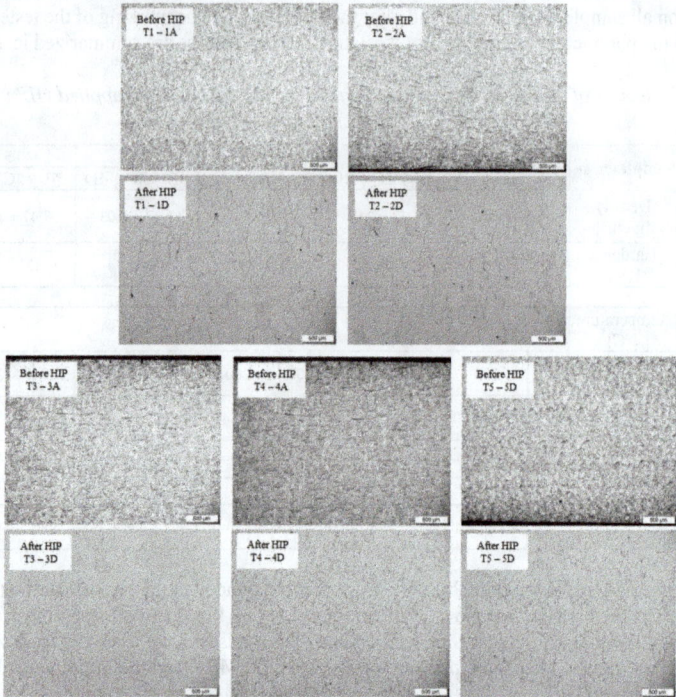

Figure 3. Samples microstructure, optical microscope, 1200°C, 200 MPa, 480 min

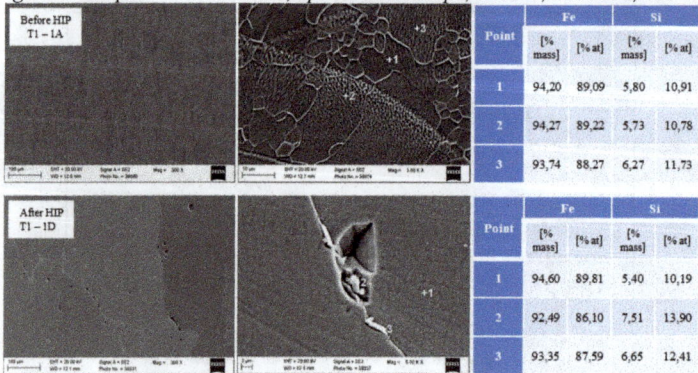

Figure 4. Microstructure of the samples, SEM

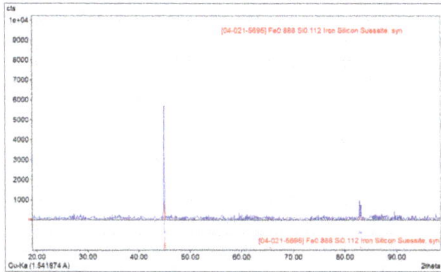

Figure 5. Phase composition of the sample after the HIP process, 1D (T1)

Discussion of the results

The analysis of the hardness and density measurement of samples after the SLM printing process shows the impact of changes in printing parameters (mainly laser power and speed) on the properties of the obtained semi-finished products. The highest density of 7.48 g / cm^3 was obtained for the sample with the T2 printing strategy, the lowest for the T1 and T3 strategies. The hardness of the printed elements was in the range of 404 - 431 HV1 (T4 and T5, respectively). Examination of the microstructure in all cases indicates the presence of irregular grains elongated in the direction of the laser beam movement (Figs. 3, 4).

After Hot Isostatic Pressing treatment, in most cases, a change in the density of prints in the range of ± 0.02 g / cm^3 (in relation to their value after SLM) was noticed depending on the SLM printing strategy and the parameters of the HIP process used. For all variants of the Hot Isostatic Pressing process (B - 1200°C, 100 MPa, 240 min; C - 1200°C, 200 MPa, 240 min; D - 1200°C, 200 MPa, 480 min), a decrease in hardness was noted, the highest by 7.66% for the sample marked with 5B, and the lowest for the sample marked with 4C by 0.99%.

Examination of the microstructure after the HIP process shows the presence of grains of different sizes depending on the printing strategy, the largest for the samples with the 1D and 2D numbers, the smallest for the samples with the 4D and 5D markings (Fig. 3). The SEM-EDS analysis showed the presence of light precipitates with increased silicon content up to 7.51 wt.%. (13.90 at. %) inside and at the grain boundaries, which, according to earlier work [6], can be related to the formation of Si-rich amorphous precipitates (Fig. 4).

Examination of the phase composition of samples after the HIP process shows the presence of a single-phase material with the chemical atomic composition of Fe - 88.8% at. and Si - 11.2% at. (Fig. 5).

The investigated microstructures show the presence of pores which, due to their regular shape, were probably formed during the preparation of samples for microstructural tests due to bright silicon-rich precipitates removed from the matrix material.

Observed herein changes in the microstructure generated by the post-treatment using the HIP method should also influence the magnetic properties of printed samples. Accordingly, optimization of HIP parameters in the context of magnetic properties will be a subject of further research.

Conclusions

As part of the research, the following conclusions were formulated:

1. The use of the Hot Isostatic Pressing process in the form of post-process treatment affects the microstructure's homogenization and improvement of printed elements' properties in SLM technology.
2. The quality of the elements obtained in the SLM printing technology impacts their

properties after using the Hot Isostatic Pressing process.

3. The combination and optimization of the SLM printing technology and the HIP process in producing FeSi6.5 magnetic soft materials has a positive effect on their properties and generates new opportunities for their use in producing elements with complex shapes.

References

[1] G. Stornelli, P. Folgarait, M. R. Ridolfi, D. Corapi, Ch. Repitsch, O. Di Pietro, A. Di Schino, Feasibility Study of Ferromagnetic Cores Fabrication by Additive Manufacturing Process, Mater. Proc., 3 28 (2021) https://doi.org/10.3390/IEC2M-09241

[2] D. Goll, D. Schuller, G. Martinek. T. Kunert, J. Schurr, C. Sinz, T. Schubert, T. Bernthaler, H. Riegel, G. Schneider, Additive manufacturing of soft materials and components, Addit. Manuf., 27 (2019) 428-439 https://doi.org/10.1016/j.addma.2019.02.021

[3] G. Stornelli, A. Faba, A. Di Schino, P. Folgarait, M. R. Ridolf, E. Cardelli. R. Montanari, Properties of Additively Manufactured Electric Steel Powder Cores Increased Si Content, *Materials* 14 1489 2021 https://doi.org/10.3390/ma14061489

[4] P. Jang, B. Lee, G Choi, Effects of annealing on the magnetic properties of Fe-6.5%Si alloy powder cores, J. Appl. Phys., 103 07E743 (2008) https://doi.org/10.1063/1.2839620

[5] D. Kołacz, K. Krukowski, A. Radoń, J. Kulasa, M. Maleta, B. Cwolek, A. Hury, Influence of the HIP process on the properties of elements obtained by additive manufacturing technologies and made in the casting process, Łukasiewicz-IMN Report, 8139 (2021), not-published.

[6] X. Shen, F. Meng, K. Boon Lau, Pei. Wang, Ch. H.T. Lee, Texture and microstructure characterizations of Fe-3.5wt%Si soft magnetic alloy fabricated via laser powder bed fusion, Mater. Character., 189 112012 (2022) https://doi.org/10.1016/j.matchar.2022.112012

Materials Research Forum LLC
https://doi.org/10.21741/9781644902837-11

Effect of Hot Isostatic Pressing on Microstructure and Properties of GH4169 Superalloy Manufactured by SLM

Shanting Niu[1,a], Hongpeng Xin[1,b], Lida Che[1,c*], Haofeng Li[1,d], Xiangyang Li[2,e]

[1]CISRI HIPEX, Haidian District, Beijing, P. R. China

[2]CISRI, Haidian District, Beijing, P. R. China

[a]niushanting@hipex.cn, [b]xinhongpeng@hipex.cn, [c]chelida@hipex.cn,
[d]lihaofeng@hipex.cn, [e]lix y@cisri.com.cn

Keywords: Hot Isostatic Pressing (HIP), Additive Manufacturing, GH4169, Properties

Abstract. Superalloy GH4169 is an extremely important material for aircraft structural components in aero-engines. In this study, the effects of different hot isostatic pressing temperature and pressure on the microstructure and mechanical properties of GH4169 were studied by means of metallographic microscope, scanning electron microscope and tensile experiment. The results show that hot isostatic pressing can significantly improve the microstructure of the test alloy manufactured by SLM. With the increase of hot isostatic pressing temperature and pressure, the tensile strength and yield strength of the test alloy decreased slightly, while the elongation showed an opposite trend. The increase of elongation is attributed to the improvement of microstructure uniformity of the test alloy by hot isostatic pressing.

1. Introduction

As the service requirements of aviation, aerospace and nuclear industry materials become more and more demanding, conventional materials have been unable to meet the requirements. At this time, the emergence of nickel-based superalloys that can maintain mechanical properties at high temperatures to meet these requirements [1]. Of nickel-based superalloy GH4169 was investigated at 650°C has high yield strength, good corrosion resistant ability and good high temperature oxidation resistance, etc., applicable to the manufacture of low temperature and under 650°C of rocket engine, aviation and ground gas turbine engine components, is currently the most widely used in aerospace field of high temperature alloy[2,3]. The alloy is a Ni-Cr-Fe base precipitation strengthening superalloy, which consists of γ matrix, γ' and γ' ' phase, δ phase and carbide[4]. The traditional manufacturing methods of GH4169 alloy are precision casting, forging and subsequent machining, etc[5]. However, this method has some problems such as low material utilization rate, low production efficiency, high processing cost and difficulty in forming three-dimensional complex structure, which greatly limits the application scope of the alloy.

Selective laser melting is one of the most widely studied and applied metal additive manufacturing technologies. Its working principle is that the metal powder is scanned layer by layer by high-energy laser beam according to the path planned by digital model slice[6]. The metal powder is melted and solidified under the action of laser energy and deposited layer by layer, thus realizing the rapid manufacturing of parts[7]. Compared with the traditional manufacturing technology, this technology has the advantages of high design freedom, one-time forming of complex parts, high material utilization rate, excellent finished product performance, and has significant advantages in preparing three-dimensional complex superalloy parts[8]. However, the results show that the GH4169 alloy samples prepared by SLM process usually contain defects such as porosity and non-fusion, and there are significant differences in the transverse and longitudinal microstructure[9,10]. There are a large number of Laves harmful phases in the micro-segregation of the molten pool solidification microstructure [11,12].

Hot Isostatic Pressing - HIP'22   Materials Research Forum LLC
Materials Research Proceedings 38 (2023) 67-77   https://doi.org/10.21741/9781644902837-11

HIP is an effective method to eliminate defects using high temperature and high pressure. Studies show that this method can effectively eliminate cracks, holes and non-fusion defects in GH4169 alloy, which has been widely concerned in the world [13]. SH Chang et al. conducted HIP treatment on Inconel718 alloy castings at different temperatures, pressures and holding time, and heat treated the corresponding samples, and then tested the tensile properties of the samples at room temperature and high temperature under different experimental conditions. The optimum HIP parameters of Inconel718 casting was determined to be 1180℃-175MPa-4h [5]. At present, there have been many studies on the HIP treatment of additive manufacturing GH4169/IN718 alloy, but the HIP process used in the present study is single, lack of studies on the influence of different HIP temperature and pressure on the microstructure and mechanical properties of SLM manufacturing GH4169 alloy. Therefore, in order to meet the requirements of engineering application, this paper studied the influence of SLM on the microstructure and mechanical properties of GH4169 alloy, and found out the more ideal process parameters of HIP.

2. Experimental

The experimental material is GH4169 alloy powder prepared by gas atomization method in electrode induction melting, and the protective gas is argon. The morphology of the powder is shown in Fig. 1, and the particle size is 15-53 μm. Fine satellite particles can be observed attached to the main particles. The particle size distribution of D10=21.9μm, D50=33.9μm, D90=52.2μm was measured by Malvern laser particle size analyzer.

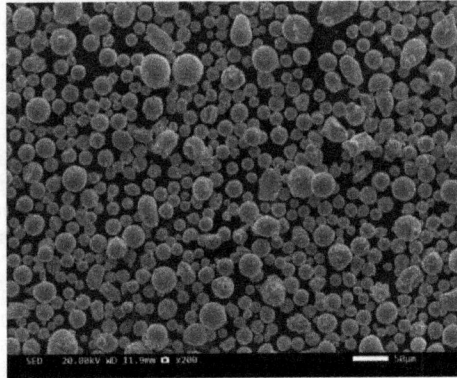

Figure 1. GH4169 powder morphology

Table 1. Chemical Composition of GH4169 alloy powder [Mass Fraction,%]

Element	Ni	Nb	Mo	Cr	Al	Ti	Co	B
Content	53.03	4.98	3.10	18.64	0.42	0.81	0.053	<0.005
Element	Si	Mn	Cu	Mg	C	S	P	Fe
Content	0.056	<0.003	0.08	<0.005	0.015	<0.003	<0.005	Bal.

In this experiment, FS271M equipment of Hua Shu High-tech was used for SLM forming of the sample, which was used to directly print the rod sample for mechanical properties testing. The sample size is φ10mm×100mm, forming direction along the sample axial vertical substrate. The main forming parameters are shown in Table 2, strip printing strategy is adopted (as shown in Fig. 2). and the printed delivery sample is shown in Fig. 3.

Table 2. Forming parameters of GH4169 alloy SLM

Laser power[W]	Scanning speed[mm·s⁻¹]	Thickness[μm]	Scanning interval[μm]
400	2700	30	80

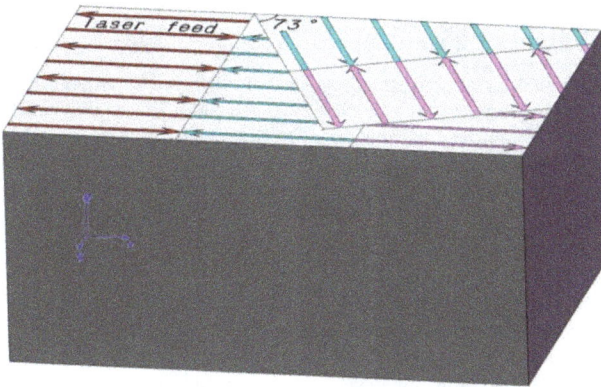

Figure 2. Schematic diagram of strip printing strategy

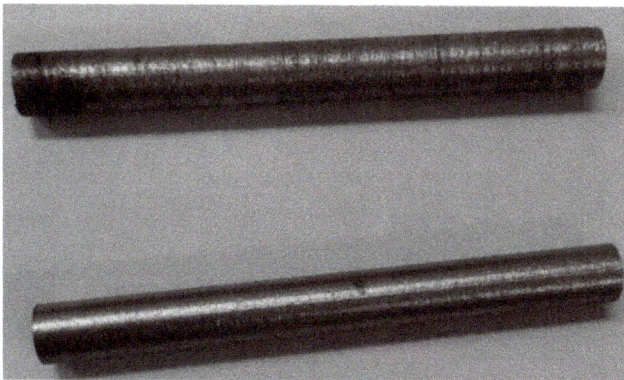

Figure 3. The printed sample

The HIP process is carried out on the sample after SLM forming. The equipment used is HIPEX80 produced by our company. The HIP parameters used are shown in Table 3 with a variety of different temperature and pressure combinations.

Table 3. HIP parameters

Number	Temp [°C]	Pressure/±5[MPa]	Soaking time[h]	Number	Temp [°C]	Pressure/±5[MPa]	Soaking time[h]
1	1100	135	4	7	1165	135	4
2	1100	155	4	8	1165	155	4
3	1100	175	4	9	1165	175	4
4	1135	135	4	10	1180	135	4
5	1135	155	4	11	1180	155	4
6	1135	175	4	12	1180	175	4

The surface morphology of polished specimens was observed by an optical microscope. The microstructures of the samples were observed by JSM-IT300LV scanning electron microscope. The microhardness of samples was measured by a microhardness tester, and 5 points were measured for each sample, and the average value was taken. The electronic universal testing machine was used to test the tensile properties at room temperature. Three samples were measured in each group and the average value was taken.

3. Results and discussion

3.1 Microstructure in GH4169 alloy

3.1.1 SLM structure and phase composition

The microstructure of GH4169 superalloy sample after SLM molding is shown in Fig. 4. Figure 4a shows the longitudinal section of the SLM sample. It can be seen from Figure 4 that the molten pool has a periodic fish scale structure. Because during the rapid scanning process, the temperature of the metal powder rises rapidly and cools rapidly, and the heat of the melt dissipates to the substrate and the surrounding powder, gradually solidifying and crystallizing to form pits. As shown in Fig. 4b, there are crisscross pool boundaries in the cross section, indicating discontinuous scanning channels.

Figure 4. Structure of SLM sample a. longitudinal section(∥BD); b. cross section(⊥BD)

In the SLM structure, dendrites exhibit epitaxial growth (Fig. 5a), with small dendrite spacing of about 0.3-1.3μm. There are a large number of fine chain phases between the dendrites, as shown in Fig. 5a. The EDS spectrum shows that the fine interdendritic precipitates are Laves phase, as shown in Fig. 5b and Table 4. The dendrite spacing is inversely proportional to the product of molten pool temperature gradient and solidification rate. Eutectic products of γ+NbC and γ+Laves can form in the interdendritic region due to severe microscopic segregation, but the number of γ+NbC eutectic is negligible compared to γ+Laves eutectic due to the very low carbon content in the material. Few other phases were observed except the Laves phase shown in Fig. 5a, presumably because rapid solidification inhibited their precipitation.

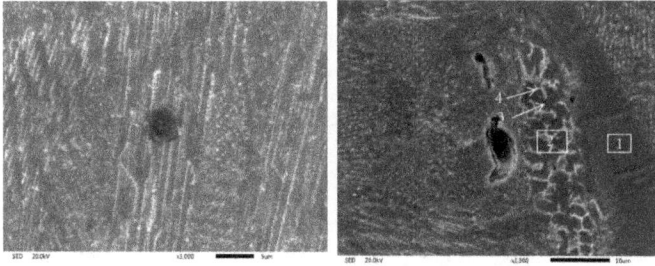

Figure 5. SLM dendrite structure

Table 4. The main chemical constituents of each phase in SLM state

Number	Phase	Ni	Cr	Fe	Nb	Mo	Ti	Al
1	γ	51.77	19.70	17.65	6.10	3.19	1.08	0.50
2	(γ+Laves) eutectic	50.51	19.45	17.61	7.25	3.69	1.08	0.41
3	γ in (γ +Laves) eutectic	52.08	20.36	19.20	3.56	3.33	0.92	0.55
4	Laves in (γ +Laves) eutectic	48.60	18.76	16.63	10.65	3.46	1.36	0.54

3.1.2 HIP structure and phase composition

Figure 6 and 7 show SEM images of cross section and longitudinal section of HIP samples respectively. At 1100°C, the microstructure of the sample still maintained columnar grain shape, and some recrystallization occurred, while some equiaxed crystals and a few twin crystals were observed. The chained Laves phase is dissolved, but the chained Laves phase and carbides are mainly distributed on grain boundaries. At 1135°C, the columnar grain morphology was almost invisible and twin crystals increased. The Laves phase is further dissolved, and most of them are granular. The Laves phase and carbides are distributed at grain boundary and in grain due to the transformation of microstructure. At 1165°C, recrystallization occurred in the sample, and a large number of twins were observed. The Laves phase is further dissolved and distributed evenly. At 1180°C, most grains of the samples grow together. The proportion of Laves phase continues to decrease, and Laves phase density areas exist locally due to grain consolidation and growth.

Materials Research Forum LLC
https://doi.org/10.21741/9781644902837-11

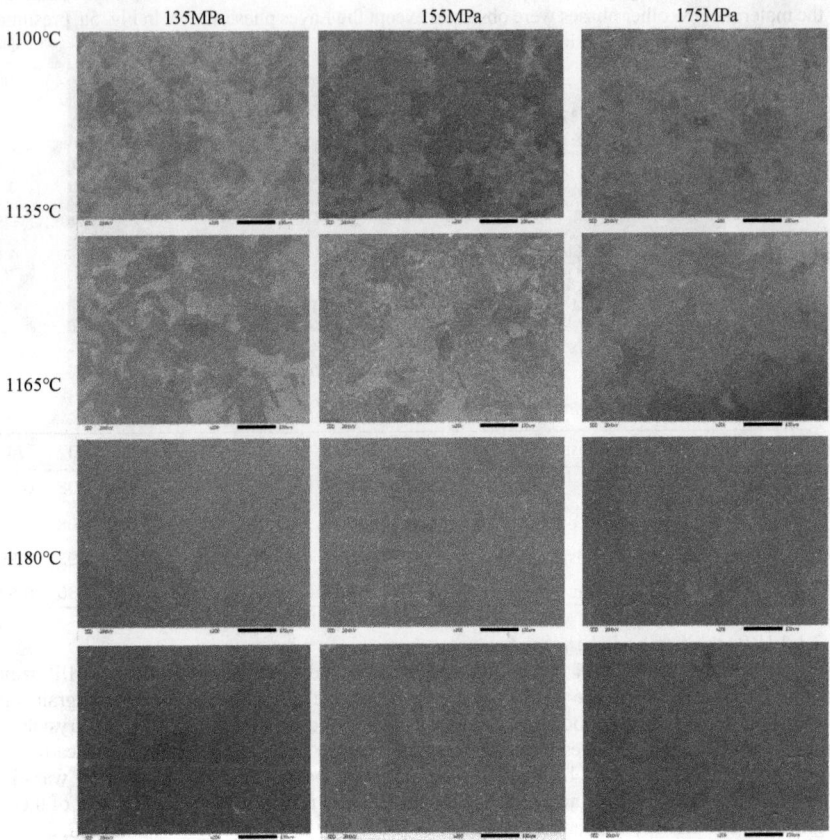

Figure 6. The microstructure of cross section (⊥ BD) of HIP specimen

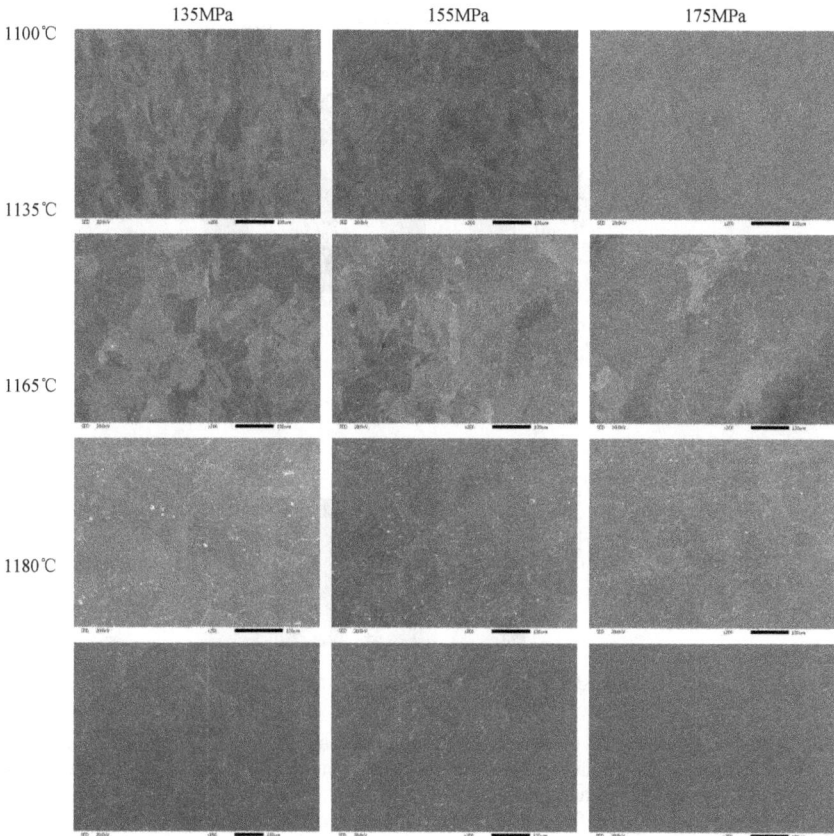

Figure 7. The microstructure of longitudinal section (∥ BD) of HIP specimen

3.2 Microhardness and density

The microhardness of samples in different states was measured by EM500-2A semi-automatic micro Vickers hardness tester. The loading force was 1kg, the pressure holding time was 15s, and the spacing between two adjacent points was 1mm. Each piece was measured 5 times on the different positions, and the average value of each position was taken. The micro Vickers hardness of the cross section (⊥BD) of the sample after different HIP is shown in Fig. 8 After SLM molding, the microhardness of the sample is 315HV1. After HIP, the Vickers hardness of the sample decreases at other temperatures except 1100°C. The strengthening mechanism of SLM alloy is solid solution strengthening and carbide and Laves phase precipitation strengthening. Due to the rapid cooling rate of SLM process, the enhanced phase cannot be precipitated. After HIP, with the increase of HIP temperature, the Laves phase gradually dissolves and the strengthening ability decreases. Meanwhile, Nb, Mo and Ti elements are released from the Laves phase and diffused into the matrix, which improves the microstructure uniformity of the material and leads to the decrease of its microhardness.

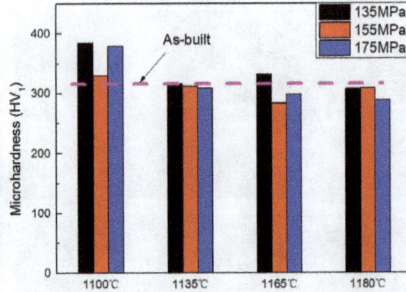

Figure 8. Microhardness of materials under different HIP treatments

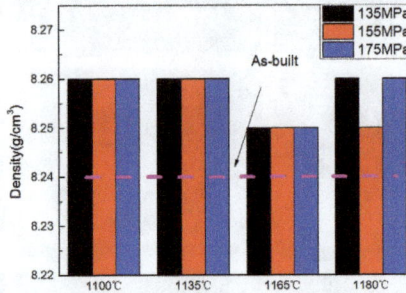

Figure 9. Material density under different HIP treatments

The density test results of the GH4169 alloy sample are shown in Fig. 9,SEM images show that the SLM sample has a small number of pores (as shown in Fig. 5), and no other defects are found. Its density is $8.24g/cm^3$, and the density after HIP is 8.25-$8.26g/cm^3$, with no obvious change in density. The density of deformed parts of GH4169 given in *handbook of aeronautical materials of china* vol.2 is $8.24g/cm^3$, indicating that the printed density is already very high, so the sample density does not increase significantly after HIP.

3.3 Tensile property

Figure 10 shows the tensile properties of SLM specimens treated by different HIP treatments. It can be seen from the figure that the tensile strength and yield strength of the specimens decrease gradually with the increase of HIP temperature, while the elongation presents an opposite trend. At 1100°C, the tensile strength and yield strength of the sample are significantly higher than those at the other three temperatures, while the elongation is significantly lower than those at the other three temperatures, which is related to the higher volume fraction of Laves phase and dislocation level of the sample at 1100°C. The strengthening mechanism of GH4169 material is solid solution strengthening and precipitation strengthening, so 1100°C should be excluded. At HIP temperature of 1135°C, tensile strength and yield strength are relatively stable under various pressures, and the rate of strength change is the lowest at 135MPa, 155MPa and 175MPa, while elongation increases at 155MPa. At HIP temperature of 1165°C, the tensile strength and yield strength of the sample at 135MPa and 155MPa were higher than those at HIP temperature of 1135°C and 1180°C. At HIP temperature of 1165C and 175MPa, the tensile strength and yield strength of the sample were between 1135°C and 1180°C, and the elongation showed a linear increase trend. At HIP temperature of 1180°C, the tensile strength and yield strength of the sample were lower than those of the other three temperatures. The tensile strength and yield strength showed a trend of rising first and then decreasing at the three pressures, while the elongation showed an opposite trend. As the cooling rate of HIP is faster than that of aging, there is almost no precipitation or strengthening phase, so the tensile strength and yield strength of the sample after HIP are lower than the forging level, but the elongation is much higher than the forging level.

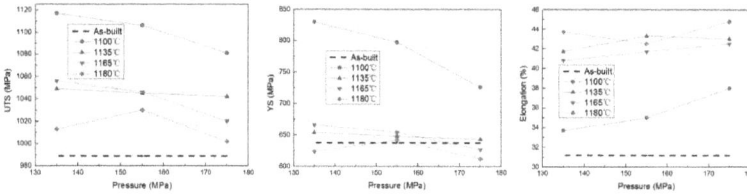

Figure 10. Tensile properties of SLM specimens treated by different HIP treatments

Figure 11 shows the tensile properties of SLM specimens treated by different HIP treatments. As shown in the figure, the tensile strength and yield strength of the specimens gradually decreased with the increase of HIP pressure, while the elongation showed an opposite trend. At HIP pressure of 175MPa, the tensile strength and yield strength of the sample continued to decrease, and were lower than those at 135MPa and 155MPa. The elongation of the sample at 1135°C was slightly lower than that at 155MPa. When the HIP pressure is 135MPa and 155MPa, the tensile strength and yield strength of the sample show a decreasing trend with the increase of HIP temperature. At 1100°C, 1135°C and 1165°C, the tensile strength and yield strength of the sample at 155MPa are lower than those at 135MPa. At 1180°C, the tensile strength of the sample at 155MPa is lower than that at 135MPa. The tensile strength and yield strength of the samples were higher than those at 135MPa. The values of tensile strength and yield strength of the samples reached the highest at 1165°C, excluding 1100°C. The elongation showed an opposite trend.

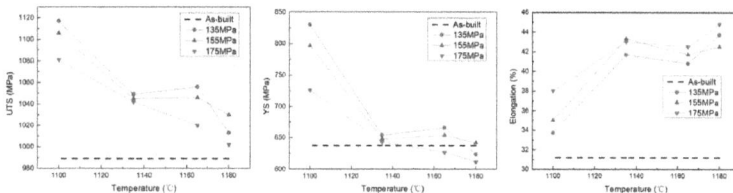

Figure 11. Tensile properties of SLM specimens treated by different HIP treatments

4. Conclusion and summary
The effects of different HIP treatments on the microstructure, SEM morphology, density, hardness and tensile properties of SLM GH4169 superalloy were investigated. The longitudinal section of SLM GH4169 superalloy is columnar grain with irregular shape, and the angle between grain and deposition direction is random. The average grain size is 10.5μm. The cross section grains are mainly equiaxed, with an average grain size of 5.93μm. In the SLM state, the dendrites are epitaxial growth with small dendrite spacing of 0.3~1.3μm. There are a lot of small chain Laves phases in the interdendrite. After HIP, the grain size of the sample increases and recrystallization occurs. Columnar grains transform into equiaxed grains on the longitudinal section. The Laves phase

gradually dissolves into granular form from the chain of SLM state. With the increase of recrystallization degree, the Laves phase is evenly distributed at the grain boundary and in the grain. There are a few pores in the SLM sample, and no other defects are found. The density of the sample is close to that of the deformed part, so the sample density does not increase significantly after HIP. With the increase of HIP temperature, the tensile strength and yield strength of the samples decreased gradually, while the elongation showed an opposite trend. With the increase of HIP pressure, the tensile strength and yield strength decrease gradually, while the elongation presents an opposite trend.

References

[1] K. Moussaoui, W. Rubio, M. Mousseigne, T. Sultan, F. Rezai.Effects of Selective Laser Melting additive manufacturing parameters of Inconel 718 on porosity, microstructure and mechanical properties, J. Materials Science and Engineering: A, 735 (2018) 182-190. https://doi.org/10.1016/j.msea.2018.08.037

[2] X. Li, J.J Shi, C.H Wang, G.H. Cao, A.M. Russell, Z.J. Zhou, C.P Li, G.F Chen. Effect of heat treatment on microstructure evolution of Inconel 718 alloy fabricated by selective laser melting, J. Journal of Alloys and Compounds, 764 (2018) 639-649. https://doi.org/10.1016/j.jallcom.2018.06.112

[3] D. Deng, R.L. Peng, H. Brodin, J. Moverare. Microstructure and mechanical properties of Inconel 718 produced by selective laser melting: Sample orientation dependence and effects of post heat treatments, J. Materials Science and Engineering: A, 713 (2018) 294-306. https://doi.org/10.1016/j.msea.2017.12.043

[4] M.T. Kim, S.Y. Chang, J.B. Won. Effect of HIP process on the micro-structural evolution of a nickel-based superalloy, J. Materials Science and Engineering: A, 441.1-2 (2016) 126-134. https://doi.org/10.1016/j.msea.2006.09.060

[5] S.H. Chang, S.C. Lee, T.P. Tang, H.H. Ho. Effects of temperature of HIP process on characteristics of Inconel 718 superalloy, J. International Journal of Cast Metals Research, 19.3 (2006) 175-180. https://doi.org/10.1179/136404606225023399

[6] K.N. Amato, S.M. Gaytan, L.E. Murr, E. Matinez, P.W Shindo, J. Hernandez, S. Collins, F. Medina. Microstructures and mechanical behavior of Inconel 718 fabricated by selective laser melting, J. Acta Materialia,60.5 (2012) 2229-2239. https://doi.org/10.1016/j.actamat.2011.12.032

[7] W. Tillmann, C. Schaak, J. Nellesen, M. Schaper, M.E. Aydinoez, K.P Hoyer. Hot Isostatic Pressing of IN718 Components Manufactured by Selective Laser Melting, J. Additive Manufacturing, 13 (2017) 93-102. https://doi.org/10.1016/j.addma.2016.11.006

[8] M. Muhammad, P. Frye, J. Simsiriwong, S. Shao, N. Shamsaei. An investigation into the effects of cyclic strain rate on the high cycle and very high cycle fatigue behaviors of wrought and additively manufactured Inconel 718, J. International Journal of Fatigue,144 (2021) 10638. https://doi.org/10.1016/j.ijfatigue.2020.106038

[9] J.N. DuPont, A.R. Marder, M.R. Notis, C.V. Robino. Solidification of Nb-bearing superalloys: Part II. Pseudoternary solidification surfaces, J. Metallurgical and Materials Transactions A, 29.11 (1998) 2797-2806. https://doi.org/10.1007/s11661-998-0320-x

[10] G.D. Janaki Ram, A. Venugopal Reddy, K. Prasad Rao, G. Madhusudhan Reddy. Control of Laves phase in Inconel 718 GTA welds with current pulsing, J. Science and technology of welding and joining, 9.5 (2004) 390-398. https://doi.org/10.1179/136217104225021788

Materials Research Forum LLC
https://doi.org/10.21741/9781644902837-11

[11] A.R. Balachandramurthi, N.R Jaladurgam, C. Kumara, T. Hansson, J.Moverare, J. Gardstam, R. pederson. On the Microstructure of Laser Beam Powder Bed Fusion Alloy 718 and Its Influence on the Low Cycle Fatigue Behaviour, J. Materials, 13.22 (2020) 5198. https://doi.org/10.3390/ma13225198

[12] Y. Kang, S. Yang, Y. Kim, B. AlMangour, K. Lee. Effect of post-treatment on the microstructure and high-temperature oxidation behaviour of additively manufactured inconel 718 alloy, J. Corrosion Science, 158 (2019) 108082. https://doi.org/10.1016/j.corsci.2019.06.030

[13] P.L. Blackwell. The mechanical and microstructural characteristics of laser-deposited IN718, J. Journal of Materials Processing Technology, 170.1-2 (2005) 240-246. https://doi.org/10.1016/j.jmatprotec.2005.05.005

Materials Research Forum LLC
https://doi.org/10.21741/9781644902837-12

Manufacturing of Net-Shape and Wear-Resistant Composite Components via the Combination of Additive Manufacturing and Hot Isostatic Pressing

Markus Mirz[1,a] *, Marie Franke-Jurisch[2,b], Anke Kaletsch[1,c], Simone Herzog[1,d]
Yuanbin Deng[1,e], Johannes Trapp[2,f], Alexander Kirchner[2,g],
Thomas Weissgärber[2,h], Christoph Broeckmann[1,i]

[1]Institute for Materials Applications in Mechanical Engineering (IWM) of RWTH Aachen University, Aachen, Germany

[2]Fraunhofer Institute for Manufacturing Technology and Advanced Materials (IFAM), Dresden, Germany

[a]m.mirz@iwm.rwth-aachen.de, [b]marie.franke-jurisch@ifam-dd.fraunhofer.de,
[c]a.kaletsch@iwm.rwth-aachen.de, [d]s.herzog@iwm.rwth-aachen.de,
[e]y.deng@iwm.rwth-aachen.de, [f]johannes.trapp@ifam-dd.fraunhofer.de,
[g]alexander.kirchner@ifam-dd.fraunhofer.de, [h]thomas.weissgaerber@ifam-dd.fraunhofer.de,
[i]c.broeckmann@iwm.rwth-aachen.de

Keywords: HIP, Hot Isostatic Pressing, Additive Manufacturing, Powder Bed Fusion – Electron Beam, PBF-EB, Composite Component, Net-Shape, Wear-Resistance

Abstract. Additive Manufacturing (AM) is an emerging technology with increasing importance in scientific and industrial applications. It is suitable for the manufacturing of very complex components straight from CAD data. Furthermore, it can complement powder metallurgical (PM) Hot Isostatic Pressing (HIP) when it is used to produce geometrical complex capsules, opposed to the manual fabrication by welding of sheet metal. This combined process route is highly automatable and can even be further enhanced when it is accompanied by numerical simulations in the design process of the near-net-shape capsules. Due to design optimization, there is no need to remove the capsule and it becomes an integral and functional part of the component. When the capsule is produced e.g., from wear-resistant materials, it can form a wear-resistant outer layer. This study comprises the manufacturing of net-shape and wear-resistant HIP capsules from the carbide rich cold working tool steel AISI A11 (X245VCrMo10-5-1) via Powder Bed Fusion – Electron Beam (PBF-EB). The capsules are filled with the tough Q+T steel AISI L6 (56NiCrMoV7) and densified by HIP with an integrated heat treatment. The focus is on the validation of the simulation, microstructural analysis, as well as analysis of the wear-resistance.

Introduction

Hot Isostatic Pressing (HIP) is a well-established powder metallurgical process for the production of highly stressed components in demanding industries. The fundamental principle of HIP is the densification of powders by applying a high pressure and high temperature regime in the range up to 200 MPa and 2000 °C [1,2]. To apply the pressure and achieve an isostatic pressure distribution over the whole surface area of the succeeding part, the powder is filled into a thin-walled capsule prior to the HIP process [2]. This is where Additive Manufacturing (AM) can complement the HIP route, when it is utilized to automatic fabricate the capsule from CAD data [3,4]. Another advantage among the automatically fabrication process is the prospect to precisely predict the shrinkage of the capsule during HIP by applying numerical simulations [5,6]. Therefore, it is possible to fabricate near-net-shape parts far more economic and ecologic with a minimal waste in material. Another advantage of the combined AM and HIP process route is the opportunity to

Hot Isostatic Pressing - HIP'22   Materials Research Forum LLC
Materials Research Proceedings 38 (2023) 78-84   https://doi.org/10.21741/9781644902837-12

use different materials for the AM fabricated capsule and the HIP densified powder. When the capsule remains on the finished part, it forms a composite component and can act as a functional layer, either corrosion- [7] or wear-resistant [8,9].

Materials and Methods

The scope of this study is to manufacture complex-shaped and fully dense composite components with a wear-resistant shell and a tough core by a combined AM and HIP process route. The HIP capsules were fabricated from the carbon-rich cold working tool steel AISI A11 (X245VCrMo105-1). An AISI 304 (X5CrNi18-10, material no. 1.4301) filling and evacuation tube was joined to the capsule by TIG welding and the capsule was filled with the Q+T steel AISI L6 (56NiCrMoV7, material no. 1.2714). The chemical composition of all materials is given in Table 1.

Table 1: Chemical composition of the HIP capsule, powder filling and filling tube in weight percent [wt.%]

Material	Part	C	Si	Mn	Cr	Mo	V	Ni	Fe
AISI A11	HIP capsule	2.4 - 2.5	0.75 - 1.10	0.35 - 0.60	4.75 - 5.75	1.10 - 1.50	9.25 - 10.25	---	Bal.
AISI L6	Powder	0.50 - 0.60	0.10 - 0.40	0.60 - 0.90	0.80 - 1.20	0.35 - 0.55	0.05 - 0.15	1.5 - 1.8	Bal.
AISI 304	Fill / Evac. Tube	< 0.03	< 1.0	< 2.0	18.0 - 20.0	-	-	8.0 - 12.0	Bal.

The AM building process of capsules for screw extruder components was conducted by Powder Bed Fusion – Electron Beam (PBF-EB), using an Arcam EBM A2X machine. An accelerating voltage of 60 kV was set and a 150 x 150 mm² build plate made from tool steel was utilized. Further process parameters were a layer thickness of 70 μm, a preheating temperature of ~ 850 °C, a line offset of 50 μm, a snake scan strategy, as well as a total area energy density of 4 J/mm². Prior to the HIP process, the remaining A11 powder within the capsule was removed and it was cleaned with alcohol. The minimum amount of powder was then calculated based on the tap density of the L6 powder and the internal volume of the capsule. The L6 powder was filled into the capsule, before it was pre-consolidated and the capsule was evacuated and closed. The subsequent densification by HIP was performed using a modernized HIP unit Shirp 20/30-2001500 from ABRA Fluid AG. The HIP parameters were chosen based on typical industrial parameters for steel powders [1] at a temperature of 1150 °C, a pressure of 100 MPa and a dwell time of 3 h. After the HIP process, the specimens were subjected to a rapid quenching within the HIP vessel. The final heat treatment by tempering was also performed within the vessel at a temperature of 540 °C for 60 minutes and repeated three times. To obtain near-net-shape components after the HIP process, a finite element (FE)-simulation has been carried out in ABAQUS to predict the final component shape and to optimize the geometry [6]. An in-house developed densification model, which considers plasticity and viscoplasticity [10,11] was implemented in a User-Material subroutine and used for the FE-simulation of the HIP process.

Results and Discussion

The first components from the combined AM and HIP process route are depicted in Fig. 1 a). The presented demonstrators are a section of a compounding extruder screw from the plastics industry. This specific part requires a very wear-resistant surface due to the constant abrasion between the screw and the plastic granulate. Additionally, the compounding process takes place under elevated temperatures up to approximately 200 °C and the extruder screw is subjected to abrupt changes in stress during start-up or when clogging occurs. This means that a very hard surface and a relatively soft and tough core are required. The deviation between the simulated and the actual measured capsule geometries after HIP is shown in Fig. 2 b). A maximum difference between 1.5 mm and ⊖○1.5 mm was measured by optical metrology, while the mean deviation was approximately 2 %. This deviation is in line with previous studies conducted on other materials by the authors [7,12,13].

Figure 1: a) Complex shaped HIP capsules manufactured by PBF-EB and b) Deviation between prediction and part

The temperature and pressure profiles of the applied HIP cycle are shown in Fig. 2 a). As the A11 capsule material is susceptible to cracking in the as-built condition, the pressure was not applied before a temperature of 700 °C. The quenching process is depicted as the blue section within the temperature profile of Fig. 2 a). The temperature was measured at the surface of the parts and the t85 time during the quenching was approximately 190 seconds. Fig. 2 b) shows the CCT diagram of A11 for an austenitizing temperature of 1030 °C. The actual measured temperature profile, plotted as the blue graph, is close to curve number 3. This curve results in a hardness of 818 HV10, while the actual measured hardness on the extruder screw was 795 HV10.

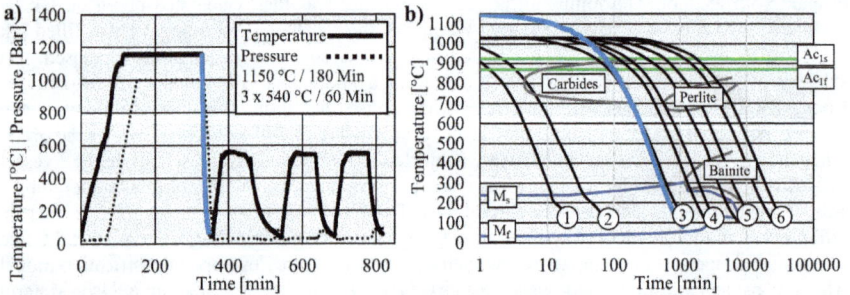

Figure 2: a) Temperature / pressure profile of applied HIP cycle b) CCT diagram of AISI A11 capsule material

This difference is plausible, since the temperature was measured on the surface of the part and the hardness was measured below the surface. Additionally, the austenitizing temperature of the extruder screw was 1150 °C opposed to the 1030 °C of the CCT diagram. However, the quenching process within the HIP vessel was successful and CCT diagrams for standard heat treatments seem to be applicable.

The microstructure of the A11/L6 composite extruder screw after HIP and tempering is depicted in Fig. 3. The bonding zone is shown at a higher magnification after etching with V2A pickle. No cracks and no pores are visible in the macroscopic picture on the left. The etched picture on the other hands shows inclusions along the bonding zone. The cause for these inclusions is not clear. However, oxide contamination of the utilized L6 powder prior to the HIP process is possibly the root cause. All aspects considered, a very good bonding between the two materials was evident.

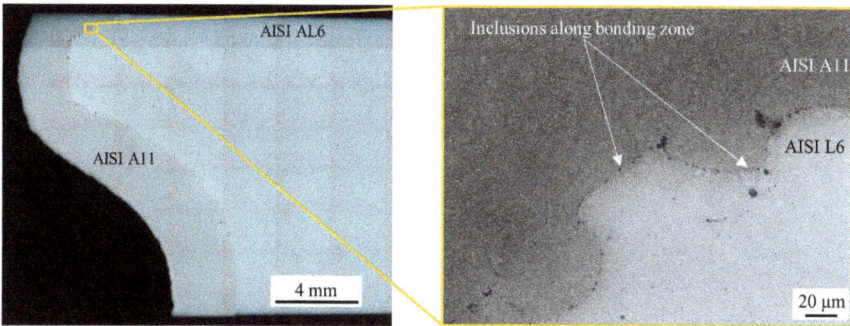

Figure 3: Microsection of AISI A11/L6 composite component and bonding zone

This finding is underlined by EDX analysis and hardness testing, as shown in Fig. 4 a) and b). An EDX line-scan across the bonding zone showed a smooth increase in vanadium and chromium content between the two materials, indicating diffusion of these elements and metallurgical bonding between the capsule and powder filling. Due to the very small and local carbide precipitates within the A11 material, the scattering for vanadium and chromium was very high within the bulk capsule. The micro-hardness profile also shows a smooth increase from ~ 460 HV0.05 for the L6 filling to ~ 750 HV0.05 at the A11 capsule.

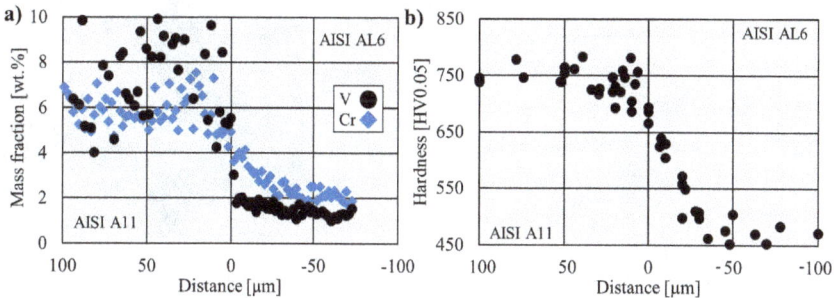

Figure 4: a) EDX line-scan along the bonding zone of A11 capsule and L6 filling indicating the local vanadium and chromium concentrations b) hardness transition between A11 capsule and L6 filling.

Figure 5 shows SEM-CBS images of the bulk A11 capsule and the bulk L6 powder filling after quenching and tempering. The microstructure of A11 consists of fine dispersed vanadium carbides within a martensitic matrix, resulting in macro-hardness of 729 ± 7.5 HV10 (~ 61 HRC). The microstructure of L6 consists of coarse and needle-shaped martensite with a macro-hardness of

481 ± 5.0 HV10 (~ 47 HRC) as a result of the triple tempering process. Inclusions are visible along the prior particle boundaries (PPB), once again indicating oxide contamination of the utilized L6 powder. The grain size distribution of A11 after HIP and quenching and tempering is depicted in Fig. 6 a) in the form of a cumulative frequency. The martensitic matrix had an average grain size of 0.28 μm, while the vanadium carbides had a coarser average grain size of 0.53 μm. Additionally, fine M_7C_3 type chromium carbides with an average grain size of 0.11 μm were detected.

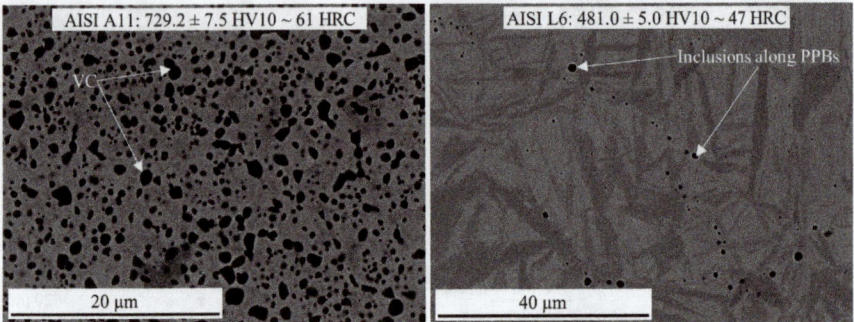

Figure 5: SEM-BSE micrographs of A11 capsule and L6 filling material.

EBSD analysis, as shown in Fig. 6 b), showed an area fraction of 22 % vanadium carbide and 1.6 % of chromium carbide. However, a distinction between carbides and retained austenite was not possible. In addition, 33 % of area fraction were not indexed confidently and the specimens were electropolished, resulting in slight topography and leading to a possible overestimation of carbide content.

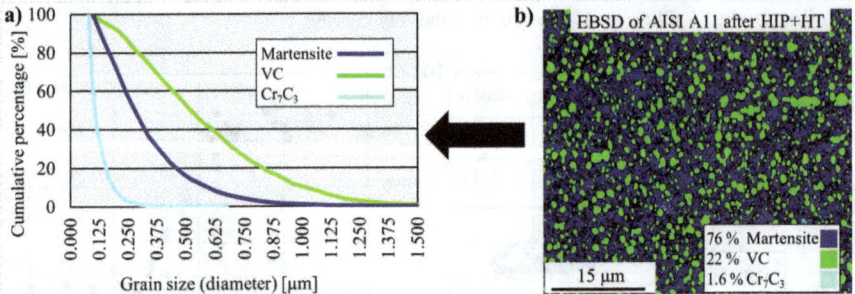

Figure 6: a) Grain size of martensitic matrix and carbides in A11 capsule material after HIP and heat treatment b) EBSD analysis of A11 capsule material after HIP and heat treatment

The wear-resistance of the A11 capsule material was characterized at Herau Anlagentechnik GmbH by a jet test, where a specimen is blasted at an angle of 15° with an abrasive medium and the volume removal per time is measured. The utilized abrasive was F80 corundum with a grain size distribution between 150 μm and 212 μm. For comparability reasons, the results were normalized to an AISI A11 cladding applied by plasma transferred arc (PTA) welding as the benchmark result. Additionally, the resistance to chipping and cracking was investigated by an impact test, indicating the toughness of the material. In this test, a weight with a ball-shaped AISI

52100 tip is dropped from different heights onto a specimen with a distance of 4 mm to the specimen's edge. The result of this test is defined as an impact energy, calculated from the drop height where no chipping or cracking occurs. Figure 7 a) and b) shows the results of the tests. Due to the high hardness of 795 HV10, the A11 PBF-EB specimen in the quenched condition outperformed the PTA cladding by 120 % in wear-resistance. Even compared to the conventionally PM-HIP produced A11 in the same heat treatment condition, the wear-resistance was significantly higher. This might be a result of the very small and evenly dispersed carbides, resulting from the rapid solidification in PBF-EB. However, the material is very brittle and showed an impact energy of only 54 J before chipping occurred. The PTA cladded A11 on the other hand reached 91 J, while the conventional PM-HIP A11 reached 81 J. After tempering, PBF-EB A11 reached an impact energy of 82 J, aligning the result co conventional A11.

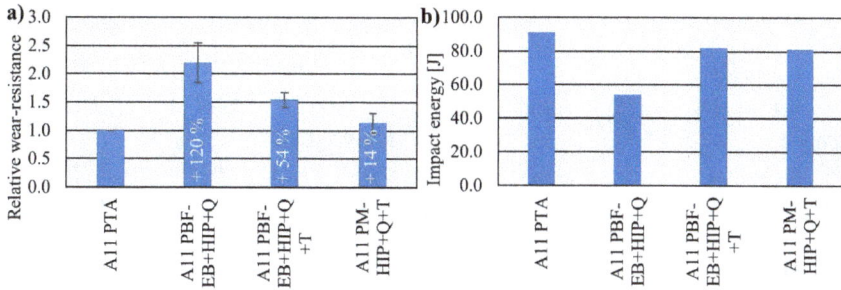

Figure 7: a) Relative wear-resistance of A11 capsule material benchmarked against PTA cladded A11 b) Resistance to chipping and cracking of A11 capsule material investigated by drop impact testing.

Summary

The feasibility of PBF-EB for the production of complex-shaped and wear-resistant capsules for HIP was proven within this study. Capsules were built from the carbide-rich cold working tool steel AISI A11 by PBF-EB and filled with AISI L6 steel powder to form a composite component. The powder was fully densified and the capsule deformation during HIP was successfully predicted by numerical simulations, leading to a net-shape component with a deviation of ~ 2 % between simulated and actual geometry. To reach a maximum wear-resistance of A11 and a high toughness of L6, the demonstrator components were heat treated within the HIP vessel straight after the densification process. In the quenched condition, A11 reached a hardness of 795 HV10, which is in line with the prediction based on CCT diagrams. This results in an outstanding wear resistance, outperforming conventional PM-HIP and PTA cladded A11. However, the material is prone to chipping and cracking in this condition. After triple tempering very small and fine dispersed chromium carbides and vanadium carbides within a martensitic matrix were observed. The hardness is lowered to 729 HV10, but the wear-resistance remains above the level of PTA cladded and conventional A11. The combined heat treatment of the wear-resistant A11 layer and the L6 filling of extruder screw component promises a high core toughness formed by a needle-shaped martensite with a hardness of ~ 481 HV10.

Funding and Acknowledgement

The research project was carried out in the framework of the industrial collective research programme (IGF no. 21074 BG). It was funded by the Federal Ministry for Economic Affairs and

Materials Research Forum LLC
https://doi.org/10.21741/9781644902837-12

Climate Action (BMWK) through the AiF (German Federation of industrial Research Associations e.V.) based on a decision taken by the German Bundestag.

References

[1]   H.V. Atkinson, S. Davies, Metall and Mat Trans A 31 (2000), 2981-3000. https://doi.org/10.1007/s11661-000-0078-2

[2]   P. Samal, J. Newkirk, Powder Metallurgy, ASM International, 2015, ISBN: 978-1-62708089-3

[3]   X. Wang, L. Carter, et al., Micromachines 11 (2020). https://doi.org/10.3390/mi11050492

[4]   S. Riehm, V. Friederici, et al., Powder Metallurgy 64 (2021) 295-307. https://doi.org/10.1080/00325899.2021.1901398

[5]   H. Hassanin, K. Essa, et al., RPJ 23 (2017) 720-726. https://doi.org/10.1108/RPJ-02-2016-0019

[6]   C. van Nguyen, Ph.D. Thesis, 2015, ISBN: 978-3-8440-4347-1

[7]   S. Riehm, A. Kaletsch, et al., in: Hot Isostatic Pressing HIP'17 (2019), Materials Research Forum LLC, 203-209. https://doi.org/10.21741/9781644900031-27

[8]   M. Mirz, B. Barthel, et al., DGM Fachtagung Werkstoffe und Additive Fertigung, 2022.

[9]   M. Franke-Jurisch, M. Mirz, et al., Materials 15 (2022). https://doi.org/10.3390/ma15051679

[10] M. Abouaf, J. L. Chenot, et al., Int J Numer. Meth Engng 25 (1988), 191-212. https://doi.org/10.1002/nme.1620250116

[11] H. A. Kuhn, C. L. Downey, Int J Powder Metall (1971), 15.

[12] Y. Deng, A. Kaletsch, et al., in: Hot Isostatic Pressing HIP'17, (2019), Materials Research Forum LLC, 182-189. https://doi.org/10.21741/9781644900031-24

[13] C. van Nguyen, Y. Deng, et al., Computer Methods in Applied Mechanics and Engineering 315 (2017) 302-315. https://doi.org/10.1016/j.cma.2016.10.033

Hot Isostatic Pressing - HIP'22
Materials Research Proceedings 38 (2023) 85-90

Materials Research Forum LLC
https://doi.org/10.21741/9781644902837-13

Development and Manufacture of Innovative Toughened Fine-Grained Recrystallized Tungsten Alloy

Koichi Niikura[1,a*], Shunsuke Makimura[2,3b], Hiroaki Kurshita[2,c], Hun-Chea Jung[1,d], Yasuhiro Matsumoto[4,e], Masashi Inotsume[4,f], Masahiro Onoi[1,g]

[1]Metal Technology Co. Ltd., Kanagawa Plant, Ebina-shi, Kanagawa, Japan

[2]Institute of Particle and Nuclear Studies, High Energy Accelerator Research Organization (KEK), Tokai-mura, Ibaraki, Japan

[3] Particle & Nuclear Physics Division, J-PARC Center, Tokai-mura, Ibaraki, Japan

[4]Sunric Co., Ltd. Fukuura, Yokohama-shi, Kanagawa, Japan

[a]kniikura@kinzoku.co.jp, [b]shunsuke.makimura@kek.jp, [c]hkurishi@post.j-parc.jp, [d]hcjung@kinzoku.co.jp, [e]matsumoto@sunric.com, [f]inotsume@sunric.com, [g]monoi@kinzoku.co.jp

Keywords: Tungsten, Recrystallization Embrittlement, Irradiation Embrittlement

Abstract. Tungsten (W) exhibits excellent thermal properties such as the highest melting point among the metallic elements, however, its engineering usefulness is limited due to the recrystallization embrittlement. Aiming at solving the recrystallization embrittlement, TFGR (Toughened Fine-Grained Recrystallized) W-1.1% TiC that exhibits high bending strength of 3.2~4.4 GPa and appreciable bend ductility at room temperature was developed. In collaboration with KEK, Sunric Co., Ltd., and Metal Technology Co. Ltd. (MTC) is upgrading the development phase for the engineering applicability of TFGR W-1.1TiC to increase the scale of manufacturing, to achieve mass production, and improve heat resistance and toughness. Reduction of gas-impurities such as oxygen and nitrogen is one of the essential factors in the manufacturing process. In this presentation, a fundamental study to understand outgassing behavior inside the HIP capsule will be introduced.

Introduction

Tungsten has the highest melting point at 3420°C among the metal elements and a high density of 19.3g/cc. These properties have been attracting attention for use in high temperature environments due to its excellent thermal expansion coefficient being the smallest as well as high thermal conductivity. However, tungsten is a brittle material exhibiting low-temperature brittleness, recrystallization brittleness, and irradiation brittleness, so the engineering applications are remarkably limited. Though it is generally well known as a filament material for incandescent bulbs, the success of the present industrial fabrication of tungsten filaments required a variety of technological developments and efforts [1]. Tungsten has also been noticed as high heat flux materials and components of fusion reactors since the 1990's and it has been considered the most suitable candidate for this project. In 2008, Kurishita et al. of Tohoku University succeeded in developing a tungsten alloy TFGR that simultaneously improved low-temperature brittleness, recrystallization brittleness, and irradiation brittleness, which were the main disadvantages of tungsten [2]. The TFGR W alloy exhibits recrystallized nanostructures containing a high density of grain boundaries (GBs) of random orientations with high energies and nanoscale precipitation and segregation of transition metal carbides such as TiCx at the GBs. The precipitation and segregation at the GBs have the beneficial effect of significantly strengthening all of the GBs, suppressing the intergranular embrittlement (low-temperature embrittlement and recrystallization embrittlement). The prototype of TFGR was about 30mm in diameter. This small size continues

Hot Isostatic Pressing - HIP'22   Materials Research Forum LLC
Materials Research Proceedings 38 (2023) 85-90   https://doi.org/10.21741/9781644902837-13

to place limits on industrial use. Makimura aimed at the application of TFGR to the large intensity proton accelerator target and the production of industrial volumes of TFGR became desired and a joint research project with MTC was established to carry out trial manufacture and mass production. Since 2017, MTC has installed a facility suitable for the process of powder preparation to sintering in order to produce TFGR industrial levels of production. The volume of toxic oxygen impurities during the process has been reduced as much as possible and a TFGR prototype was completed in 2019.[3] At present, the development of tungsten alloy is being promoted by constructing a project system with Sunric Co., Ltd. aiming at use in a wider array of industrial fields. This paper summarizes the present state of evaluating the oxygen concentration reduction measures. Oxygen concentration reduction measures are one of the most important points in the preparation of TFGR tungsten alloy.

Evaluation of oxygen concentration

TFGR W-TiC, which exhibits appreciable bend ductility even at room temperature, is manufactured in the following process such as; 1. Purification of powder without contact with oxygen and nitrogen during handling of the powder, 2. Mechanical alloying (MA), which is carried out in a vacuum vessel with balls with W-TiC powder to refine crystal grains of W-TiC to W matrix, sintering at a low temperature as possible to keep the grain-size refined, 3. Microstructural modification by grain boundary sliding (GSMM: Grain boundary Sliding-based Microstructural Modification) to promote grain boundary segregation and precipitation of TiC in the W matrix [2]. At the time of implementing all the measures examined as the measure of the O2 level, the oxygen level after hot isostatic pressing, was 240ppm (wt. ppm). In order to reduce the DBTT (Ductile-Brittle Transition Temperature) in the three-point bending test to room temperature or lower, the target oxygen concentration is 450ppm or lower.

Degassing equipment

Mixing and degassing of W powder and TiC powder, loading and unloading of powder into and out of the MA container, filling the HIP capsule, and welding are all carried out inside the glove box (GB). The GB is filled with an inert gas, argon (Ar), and the oxygen concentration is kept at 1ppm or less by an Ar gas circulation purifier at all times. A vacuum chamber for heating and degassing is attached to the side of the glove box. A special welding machine is connected to weld the HIP capsules (Fig.1). The vacuum chamber for heating/ degassing is supplied with a door for manual access from within the GB. A turbo molecular pump, TMP, and a dry pump are used for the vacuum exhaust and a quadrupole mass spectrometer, Q mass, is mounted for analysis of the exhausted components.

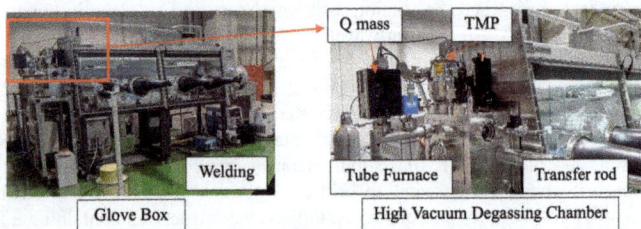

Figure 1. Pictures of the glove box: OMNI-LAB (VAC), vacuum chamber, a TMP pump: TG350F (OHSAKA VAC), a tube furnace: TMF700N (AZONE) and a Q mass: HPS2 (MSK).

Pass boxes are used for taking objects in and out of the GB while the atmosphere is replaced by Ar gas in a vacuum and the low oxygen state is always maintained. The vent line of the vacuum chamber uses Ar commonly circulated in the GB. The overall system diagram is shown in Fig. 2.

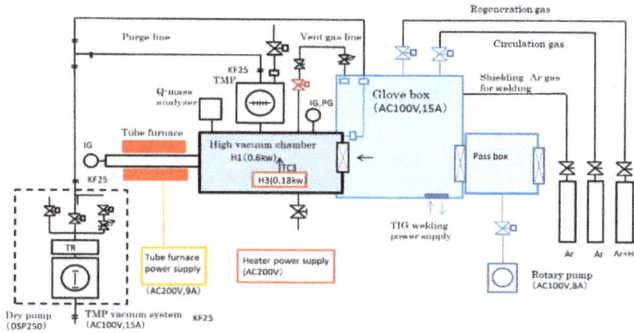

Figure 2. Schematic Diagram of GB, High Vacuum Chamber, Tube furnace

Degassing

Mixing of the powders is carried out in the GB using an electronic balance. The blended powder is placed in a tantalum, Ta boat and attached manually to the tip of the high vacuum chamber in an Ar atmosphere in the GB. The powder is evacuated from the chamber and moved to the vacuum tube furnace by a transfer rod. Degassing is carried out at 950°C for 1 hour. After degassing, only the powder is placed into the MA vessel containing the balls for MA which is in a high vacuum chamber. The lid of the MA container attached to a vertical drive shaft is lowered with the transfer rod, then the vacuum chamber is vented, and the container is sealed by vacuum load against atmospheric pressure. Finally, bolts are inserted into the through-holes of the container in the GB. The heated degassing and high vacuum chambers can be seen in Fig. 3.

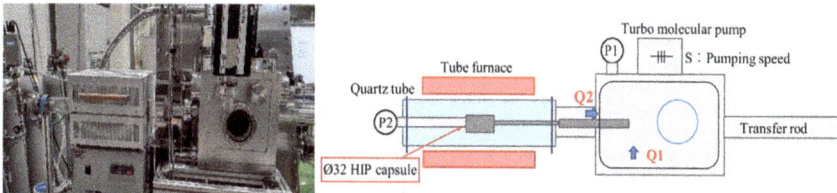

Figure 3. Photographs and Block Diagrams of Heated Degassing and High Vacuum Chambers

HIP Capsule Welding/ Degassing Treatment

The MA powder is packed in a HIP capsule in the GB, then the capsule with a pipe is enclosed through TIG welding the HIP capsule and is placed under a vacuum while heated to eliminate residual gas. Ar gas that creates an inert atmosphere to ensure a stable TIG discharge is supplied separately. A leak test is performed after welding using the exhaust port in the GB. Thereafter, the HIP capsule and the powder is heated and degassed in the tube furnace. Finally, the pipe for pumping is pinched and welded to realize sealing. A schematic of the HIP capsule can be seen in Fig. 4.

C4：HIP capsule
Ø30×60mm
C2 : Steel pipe
Ø6×240mm

C3 : Orifice
C1: SS flexible tube
Ø12.7×460mm

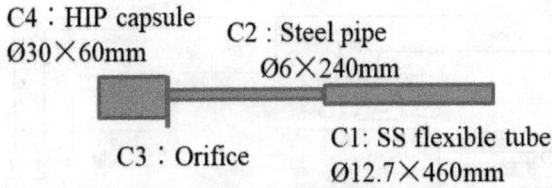

Figure 4. HIP-Capsule and Pipe Schematic Drawing

Conductance of powder inside HIP capsule

So far, it was verified that the powder filling and degassing in the HIP capsule were sufficiently carried out. When the capsule is pumped through the pipe, the vacuum pressure at the bottom of the HIP vessel depends on the conductance of the powder. According to previous literature [5], the conductance through the powder has been measured for viscous and molecular flows in a relatively low vacuum. The conductance depends on the vacuum pressure, and in both cases, it is proportional to the cross-sectional area of the pumping path and inversely proportional to its length. In this study, the vacuum pressures at the bottom of the HIP vessel and at the vacuum pump position were measured to estimate the amount of degassing and the conductance in high vacuum. Degassing rate of the stainless steel and the quartz can be assumed from the handbook of the vacuum science. The amount of total degassing can be estimated by multiplying the surface area of the vacuum chamber of the stainless steel and the surface area of the quartz tube furnace. From this degassing quantity and the pumping speed from the turbo molecular pump, the vacuum pressure was predicted and the comparison with the measured value was carried out (Table 1). Furthermore, the conductance can be estimated from the diameter and length of the piping and the HIP capsule. As the result, it was confirmed that the prediction is consistent with the measurements for the empty HIP capsule that does not contain powder (Table 2).

Table 1. Confirm of Attained Vacuum

Pumping speed of turbo pump	310	L/sec
S: Effective pumping speed	0.295	m^3/s
Q1: Outgassing from chamber	9.94×10^{-6}	Pam^3/s
Q2: Outgassing from tube furnace	9.15×10^{-7}	Pam^3/s
Ultimate vacuum (Q1+Q2)/S	3.68×10^{-5}	Pa
Measured value	4.50×10^{-5}	Pa

Calculated Pressure	Measured Pressure
P1: 3.74×10^{-5} Pa	P1: 4.5×10^{-5} Pa
P2: 4.22×10^{-3} Pa	P2: 8.5×10^{-3} Pa

Materials Research Forum LLC
https://doi.org/10.21741/9781644902837-13

Table 2. Conductance and outgassing of HIP capsule + pipe.

Conductance

C1: SS flexible tube	5.39×10^{-4} m³/s
C2: Ø6 Steel pipe	3.23×10^{-5} m³/s
C3: Orifice	1.16×10^{-4} m³/s
C4: HIP capsule	5.45×10^{-2} m³/s
C capsule: Total conductance	2.41×10^{-5} m³/s

Outgassing

q1: SS flexible tube	4.04×10^{-10} m³/s
q2: Ø6 Steel pipe	3.32×10^{-8} m³/s
q3: HIP capsule	6.74×10^{-8} m³/s
qT: Total outgassing	1.01×10^{-7} m³/s

The amount of outgassing from the powder was estimated from the vacuum pressure P1 near the turbo-molecular pump and the pumping speed by putting the powder into the HIP capsule afterwards. Then, the conductance was determined from the differential pressure of the vacuum pressure P2 at the bottom of the HIP capsule and P1. A quadrupole mass spectrometer, Q mass, was used to identify the types of gas released when the HIP vessel with a diameter of 32 mm and the powder were degassed (Fig. 5).

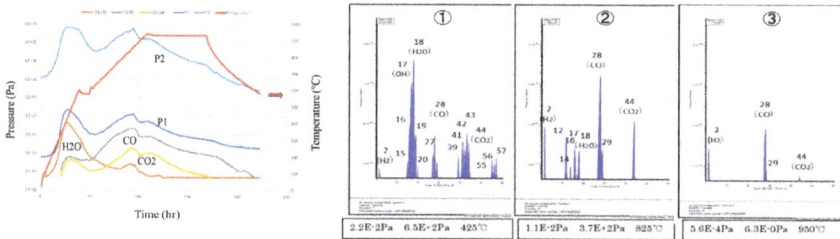

Figure 5. Degassed Properties with Powder and Q mass Spectral Data at Elapsed Time Points ①②③

From the differential pressure, the conductance after degassing of the powder was calculated that at a diameter of 32mm capsule was 2.56×10^{-4} Pam3/s (Table 3). It was confirmed that the HIP capsule bottom was exhausted to 1.3×10^{-3} Pa even when the HIP capsule was considerably packed with powder.

Table 3. Powdered Vacuum and Powder Conductance Calculations

P1(measured)	1.60×10^{-4}	Pa
P2(measured)	1.30×10^{-1}	Pa
Outgassing $Qt = P1 \times S$	4.72×10^{-5}	Pam³/s
Amount of gas other than powder Q0	1.10×10^{-5}	Pam³/s
Amount of gas from powder $Qp = Qt-Q0$	3.62×10^{-5}	Pam³/s
Outgassing from HIP capsule and pipe Qh	1.01×10^{-7}	Pam³/s
Conductance of powder $(Qp+Qh)/(P2-P1)-Ct$	2.56×10^{-4}	m³/s

Discussion

It was verified that the specifications required for the fabrication of TFGR W alloys were satisfied in the present pumping system. Subsequently, the largest dimension that can be introduced into the tube furnace is 50mm in diameter. It was confirmed that the conductance is 8.66×10^{-4} Pam^3/s.

Degassing rate of W powder. The degassing rate of W powder was calculated from the total amount of outgassing and powder from the measured W powder of 200g. Assuming the powder particle diameter as a sphere of 2μm, the weight of one particle is determined to be 8.0×10^{-11}g. The volume of W powder in the HIP capsule is 200g, then the number is 2.49×10^{12}. The surface area of the particle is 1.26×10^{-7}cm^2 assuming a spherical shape, and the total surface area is 9.42cm^2. Dividing the degassing amount of the powder 3.5×10^{-5}Pam3/s by the surface area becomes 1.12×10^{-6}Pam^3/s/m^2. This value is close to the normal metallic surface (e.g., SS304) and looks reasonable.

Conductance of gas permeating W powder. The conductance of the powder–filled in HIP capsule was calculated to be 2.56×10^{-4}m^3/s. For comparison, we calculated the diameter to be 4.7mm if this conductance were in the shape of a pipe and length were 50mm. This diameter was larger than expected. It is presumed that the reason for this result was that the powder had a large conductance. As a results, the pressure of the bottom of the HIP capsule was lower than expected.

Summary

Though the production of TFGR W alloy prepared until now by MTC is not yet at a mass production level, it was judged that the basic parameters required for manufacturing larger sizes has been confirmed.

- Treatment of all powder handling without touching the atmosphere resulted in a 240ppm oxygen-concentration for TFGR W alloy fabrication (must be less than 450ppm) [4].
- The vacuum at the bottom of the HIP capsule reached 1.3×10^{-1} Pa using the degassing process for the HIP capsule even when the W powder was packed. When the pressure of P1 entrance of HIP capsule was 1.60×10^{-4}Pa which was two orders of magnitude smaller than the bottom value.
- Analysis of the Q mass in the degassing step of the W powder at 950℃, it was confirmed that the peak of the CO is present even after 1 hour has elapsed. In this degassing system, it is considered that the oxygen concentration is further improved by treatment at high temperature for a longer period of time.

References

[1] T.Igarashi Materia Japan 42 (2003)533-538. https://doi.org/10.2320/materia.42.533

[2] H.Kurishita et al., Mater. Trans. 54 (2013) 456-465. https://doi.org/10.2320/matertrans.MG201209

[3] S.Makimura et al., Mater.Sci.Forum,Spallation Materials Technology,1024(2021)103-109. https://doi.org/10.4028/www.scientific.net/MSF.1024.103

[4] H.Kurishita et al.,Phys.Scr.T159(2014)012032(7pp). https://doi.org/10.1088/0031-8949/2014/T159/014032

[5] H. Nakagawa et al.,- Vacuum Vol. 5, No. 2 (1961) 19-25.

Hot Isostatic Pressing - HIP'22   Materials Research Forum LLC
Materials Research Proceedings 38 (2023) 91-100   https://doi.org/10.21741/9781644902837-14

Comparison of HIP Composite and HIP Solid Material with Melting Metallurgically Produced Solid Material

Alexander Ernst[1,a*], Beat Hofer[2,b], Adem Altay[1,c], Michael Hamentgen[1,d]

[1]Saar-Pulvermetall GmbH, Werner-von-Siemens-Straße 23, 66793 Saarwellingen, Germany

[2]Hofer Werkstoff Marketing Beratung, Steinmattstraße 25, 4552 Derendingen, Switzerland

[a]alexander.ernst@saar-pulvermetall.de, [b]beat.hofer@hoferwmb.ch,
[c]adem.altay@saarpulvermetall.de, [d]michael.hamentgen@saar-pulvermetall.de

Keywords: Hot Isostatic Pressing, Powder Metallurgy, Melting Metallurgy, Composite Material, HSS, Twin Screw Extruders, Rollers, Pellet Mills, Mechanical Properties, Quality Improvement, Service Life Extension

Abstract. The production of composite materials using Hot Isostatic Pressing (HIP) technology is finding more and more applications. Composite parts for co-rotating twin screw extruders have been introduced in practice for years and are indispensable. Further applications in the roller sector for the processing of high-quality sheet metal and wood pellet production contribute to quality improvement and service life extension and can be tailored precisely to the demands. Composite material is particularly advantageous for large dimensions. Requirements such as high hardness in the rim zone and high strength in the load-bearing core can be combined with different materials. In the present work, HIP composite materials and HIP solid materials were compared with melting metallurgically produced solid materials. The differences in hardness and strength in the rim zone and in the core after heat treatment are presented, showing the advantages and disadvantages of both systems. The choice of composite material depends on the requirements and can cover contrary needs such as corrosion, wear resistance and high strength.

Introduction

Hot Isostatic Pressing (HIP) was developed 1955 at the *Battelle Memorial Institute's Columbus Laboratories* in Columbus (Ohio, USA) initially for cladding nuclear fuel elements for submarines [1,2]. Since those days, HIP has established itself as a reliable manufacturing process for complex and highly stressed components in different industrial sectors, e.g., for the aerospace, offshore, mechanical engineering or energy sector [3,4].

As specialty, the diffusion processes taking place during HIP are used to produce composite parts and thus create tailored properties at those points of a tool where they are needed. In twin screw extruders for plastic processing, for example, the screws are made by a combination of a conventional tempering or tool steel with a high toughness in the core and a corrosion and/or abrasion-resistant powder metallurgical (PM) material in the rim zone. As a result, the tough core allows operation at high torques and thus ensures an increase in productivity while the PM layer enables service life to be extended by up to a factor of 10, depending on the chosen material and the materials to be processed [5].

In the production of metal sheets, highest requirements are placed on the surface integrity of the working rolls, since irregularities on the roll surface are transferred directly to the sheet, where they lead to intolerable deviations in shape and dimensions. To maintain surface integrity as long as possible, the use of suitable materials is essential. PM high-speed steels (HSS) are ideal for this purpose, as they have a special microstructural design leading to high wear resistance and, depending on their chemical composition, high corrosion resistance, and can therefore meet the demands in cold rolling mills to the highest degree. In this case as well, a composite with a tough

Materials Research Forum LLC
https://doi.org/10.21741/9781644902837-14

core is used to absorb the alternating loads during the rolling process and to assure the longest possible service life of the rolls. Moreover, productivity can be improved by applying higher torques.

In order to withstand the mineral and fiber content of the wood, acting like lubricating paper to pellet rollers and pan grinders, a HIP composite solution can be used. Here, also, a combination of two different requirements have to be met. Hard on the outside, solid and tough on the inside. This ensures to improve wear resistance while absorbing the high pressing forces. Considerable service life improvements by a factor of 3 and more have been achieved [5].

Now to demonstrate and compare the special properties of a HIP composite to solid material parts, in this study three bars, namely a solid bar made of PM HSS, a composite bar of PM HSS/tool steel and a solid bar of conventional HSS are produced and hardened. The advantages and disadvantages of the different systems are shown by comparing the transverse rupture strength (TRS) and hardness profiles of the components.

Materials and Experimental Setup

The parts studied in this contribution consist of the following materials:

1) Solid bar made of powder metallurgically produced HSS *PM 1.3344* (SARAMET 23, AISI M 3-2 PM)
2) Composite bar made of *PM 1.3344* in the rim zone and melting metallurgically (MM) produced tool steel *1.2344* (AISI H13) in the core.
3) Solid bar made of melting metallurgically produced HSS *MM 1.3344* (AISI M 3-2).

The chemical compositions of the used materials are presented in Table 1. The differences between PM 1.3344 and MM 1.3344 lie in the microstructure, particularly in the grain size as well as in the size and distribution of the carbides, with PM 1.3344 having a significantly finer-grained microstructure in which numerous small and homogeneously distributed carbides are embedded in the matrix, c.f. Fig. 1 (a). In contrast, the conventional production route via ingot casting causes considerably bigger grains as well as coarse and linearly arranged carbides in MM 1.3344, which is illustrated in Fig. 1 (b). This microstructural texture can be further intensified by the following machining steps, while the HIP process ensures an isotropic microstructure in PM 1.3344 [6].

The bars were produced in the dimensions Ø 200 x 500 mm, c.f. Fig. 2. In the case of the composite bar, the capsule was first fitted with a solid bar of 1.2344 with a diameter of 140 mm, and then filled with PM 1.3344 powder. After mechanical pre-densification, sealing and evacuation, the PM 1.3344 and the composite bar capsules were hipped together in one load with the parameters 1100 bar/1150 °C/4 h/cooling 40 °C/h. After checking the argon content [7] and release of the manufactured bars, the capsule material was mechanically removed via turning. The MM 1.3344 bar and the 1.2344 core bar were purchased commercially in the rolled and machined condition.

Subsequent to the HIP process a heat treatment under vacuum atmosphere was conducted. All three bars were subjected to the following heat treatment in one furnace load:

- Hardening: 1130 °C/45 min/pressurized nitrogen cooling down to 350 °C/100 min holding to enable temperature balance between rim zone and core/pressurized nitrogen cooling to RT

Hot Isostatic Pressing - HIP'22   Materials Research Forum LLC
Materials Research Proceedings 38 (2023) 91-100   https://doi.org/10.21741/9781644902837-14

Table 1: Chemical compositions of the used materials listing the main alloying elements in wt.%

Material	C [%]	Si [%]	Cr [%]	Mo [%]	W [%]	V [%]	Fe [%]
PM 1.3344	1.3	0.5	4.2	5.0	6.4	3.1	Bal.
1.2344	0.4	1.0	5.3	1.4	-	1.0	Bal.
MM 1.3344	1.3	0.3	4.0	5.0	6.2	3.0	Bal.

- Tempering: 530 °C/4 h/air cooling
             500 °C/4 h/air cooling
             500 °C/4 h/air cooling

After heat treatment a plate with a height of 35.2 mm was cut out of the center of each bar via wire-EDM, shown exemplarily for the composite bar in Fig. 2. During EDM, the MM 1.3344 bar cracked, see Fig. 3, which is a result of massive residual stresses introduced into the bar by the hardening process. However, the crack doesn't influence the mechanical testing of this bar. In contrast, the PM 1.3344 bar and the composite bar could be machined without any problems. Due to the fine-grained, segregation-free microstructure of the PM material, the microstructural transformation and precipitation processes during heat treatment occur much more homogeneously than in the MM material, resulting in significantly lower hardening stresses.

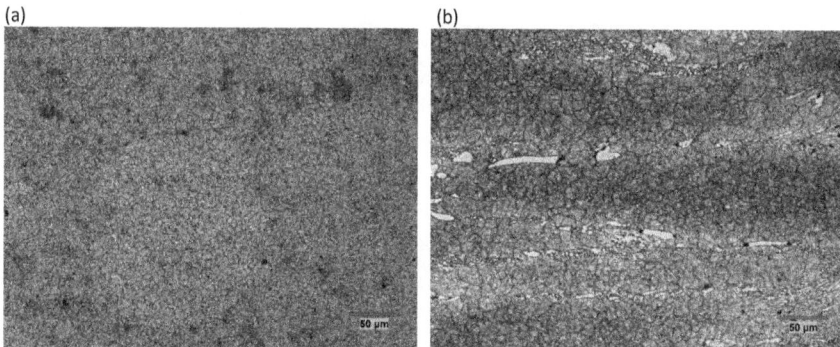

Figure 1. Micrographs showing a fine-grained microstructure with numerous small, homogeneously distributed carbides in PM 1.3344 (a) and bigger grains with coarse, linearly arranged carbides in MM 1.3344 (b)

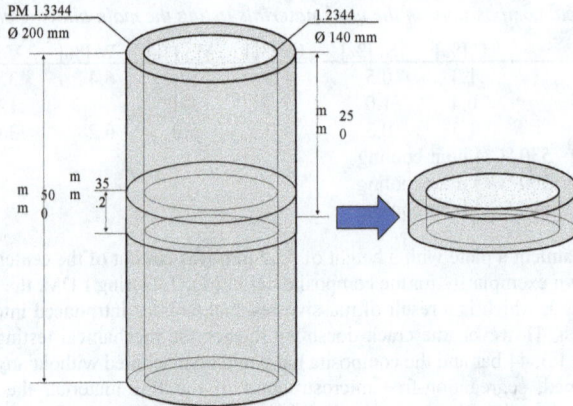

Figure 2. Bar dimensions and plate extraction out of the center, exemplarily shown for the composite bar.

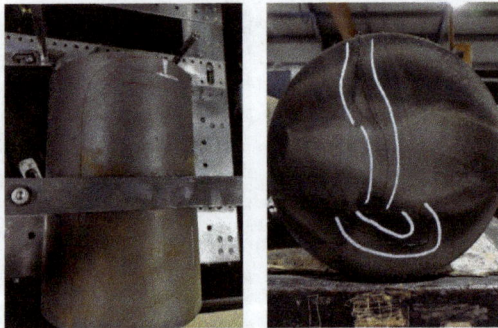

Figure 3. The stresses introduced by the heat treatment cause the MM 1.3344 bar to crack during EDM.

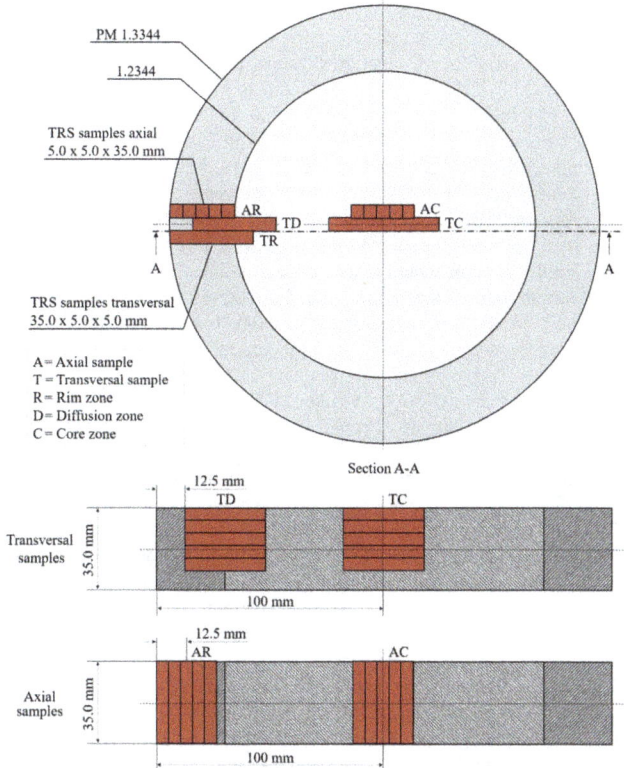

Figure 4. Sampling plan illustrated by the example of the composite bar PM 1.3344/1.2344

In order to measure the hardness profile, both in x- and y-direction six hardness measurements were performed close to the edge and every 10 mm over the cross-section of the components using an *Equotip 550* mobile hardness tester.

To check the strength characteristics of the three bars, TRS tests were performed acc. to DIN EN ISO 3327 [8]. Axial (A) and transversal (T) samples were eroded out of different depths, from the rim zone (R) over the diffusion zone (D) to the core (C), and ground on a surface grinding machine. The sampling plan in Fig. 4 shows that 25 samples were manufactured out of each plate.

Results and Discussion

The hardness measurement results over the cross-section are presented in Fig. 5. The mean values at each measuring depth are depicted in the form of columns. The measurement scatter is recorded on basis of the standard deviation indicated by error bars.

The PM 1.3344 solid bar, Fig. 5 (a), exhibits a homogeneous hardness profile up to the component core. Over all measuring depths the hardness values vary between 61.6 and 63.6 HRC. This small scattering is an indication of a homogeneous hardening microstructure, which results in uniform mechanical properties over the entire cross-section of the bar.

The results in Fig. 5 (b) show that the composite bar reveals a hardness of 62.1 - 63.1 HRC in the PM 1.3344 layer up to a measuring depth of 30 mm, which is on the same level as the PM

1.3344 solid bar. All measurements above 30 mm are made within the range of the conventional tool steel 1.2344. Here the hardness values experience a significant drop to 57.0 HRC and fall down to 55.3 HRC with increasing distance from the surface.

In the case of the MM 1.3344 solid bar, presented in Fig. 5 (c), the hardness values are at a comparable level to PM 1.3344. However, it can be seen that the hardness profile scatters more widely with values between 59.5 - 63.1 HRC. Additionally, the measurements within the individual measuring depths also scatter more widely, caused by microstructural inhomogeneities coming from the melting metallurgical manufacturing process.

The TRS test results are illustrated in Fig. 6. Each column represents the mean value of five measurements on every sample. The standard deviation is featured again by error bars. In the case of the PM 1.3344 solid bar, represented by the blue columns, the TRS shows no dependence on the location or the orientation of the sample. The hot isostatic compaction of the metal powder ensures a fine-grained, homogeneous microstructure, which allows uniform hardening right into the component core on the one hand, and isotropic material properties on the other hand, so that the values for transversal and axial samples are at the same level. In addition to high absolute values, this results in an extremely low scattering of the hardness values of less than 3 %.

The results of the composite bar are depicted in orange columns in Fig. 6. In the zone TD, the samples are positioned in such way that the test load is applied at the boundary between PM 1.3344 and 1.2344 (diffusion zone). All samples of this section did not break in the diffusion zone but in the region of the 1.2344 bulk material. Thus, at TD, the column characterizes the TRS of the high toughness 1.2344 tool steel, which achieves comparable values to PM 1.3344. The samples at positions TR and AR are located almost completely within the range of PM 1.3344 and show therefore similar measurement results to the PM 1.3344 solid bar. The samples from positions TC and AC lie entirely in the range of 1.2344. The measurement results at position TC are about 8 % below the values of TD. This could indicate that the grains get coarser from the rim to the core because of different deformation ratios during the rolling process, which is explained more in detail for the MM 1.3344 bar in the following section. The axial samples at this depth AC provide about 3% more in TRS compared to the transversal samples TC. This can be explained by the microstructural texture resulting from the manufacturing process of the bar. During rolling, the grains get stretched in rolling direction which leads to the formation of the so-called "forging-fiber" that runs through the core bar in axial direction. As a result, during the TRS test, the axial samples are loaded perpendicular to the forging fiber and can therefore offer greater resistance to breaking than the transversal samples which are loaded parallel to the forging fiber. This influence is reduced by coarser grains in the core area, resulting in this slight difference of only 3%.

(a)

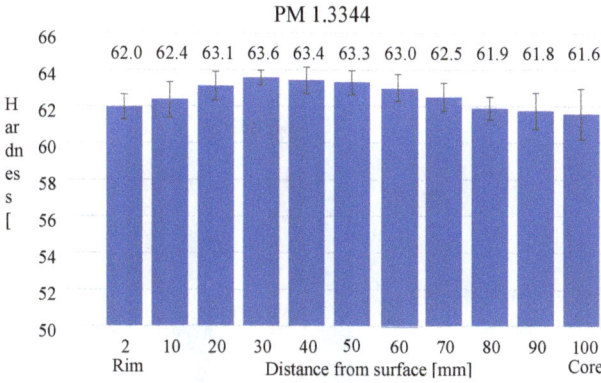

(b)

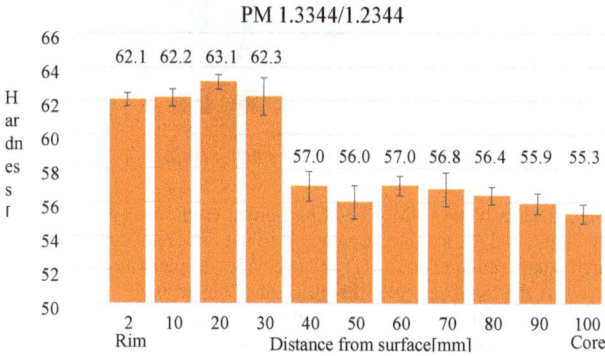

(c)

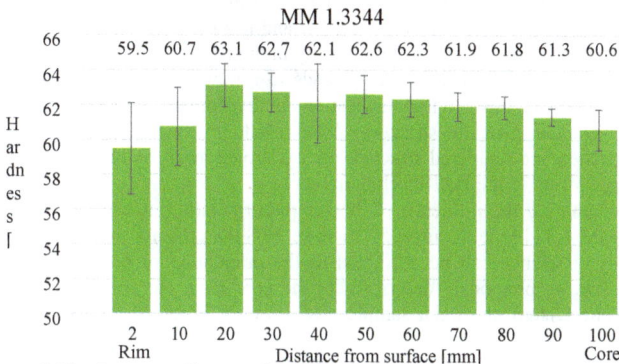

Figure 5. Hardness profile over the cross-section for PM 1.3344 solid bar (a), PM 1.3344/1.2344 composite bar (b) and MM 1.3344 solid bar (c). At each measuring depth 12 measurements were performed using a mobile hardness tester

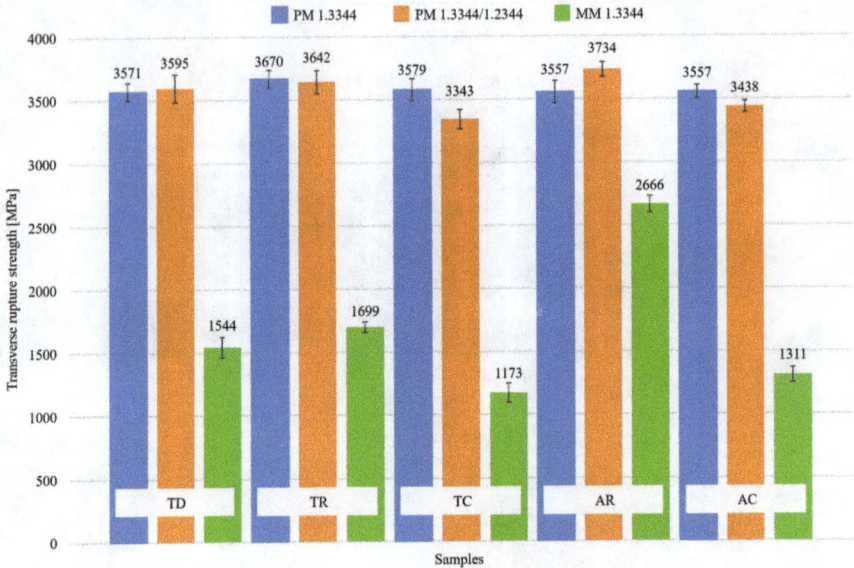

Figure 6. TRS measurement results

The measurement results of the MM 1.3344 bar, green columns in Fig 6, show a significant dependence on the depth and the sample orientation. In the rim zone at position TR, 10 % higher values are measured compared to position TD and even 45 % higher values compared to TC in the bar core. In the case of the axial samples, this influence is even more evident. Here, double the values are measured in the rim zone AR compared to the core AC. As mentioned in the section above, this fact can be explained by differences in grain size between rim and core, c.f. Fig. 7. During the production of the bar, the rolling process causes a high deformation level in the area close to the rim. This deformation decreases further and further on the way to the core. The high deformation level ensures a high dislocation density in the rim zone. During the subsequent heat treatment, a high dislocation density initially leads to a high number of sub-grain boundaries (recovery) which in turn results in a high nucleation number for new grains (recrystallization) and finally in a fine-grained austenite structure. As a result, the fine-grained austenite enables a fine-grained quenched and tempered structure after completion of the heat treatment and thus a higher strength according to the Hall-Patch correlation.

In addition, examining the influence of the sample orientation, it can be seen that the axial samples AR of MM 1.3344 yield almost 57 % more TRS than the transversal samples TR at the same depth. This difference can be attributed to the orientation of the forging fiber in axial direction of the bar, as previously described. In the core area, the coarser grains weaken the orientation influence. Here, the axial samples AC lie only about 12 % above the transversal samples TC.

Materials Research Forum LLC
https://doi.org/10.21741/9781644902837-14

(a)

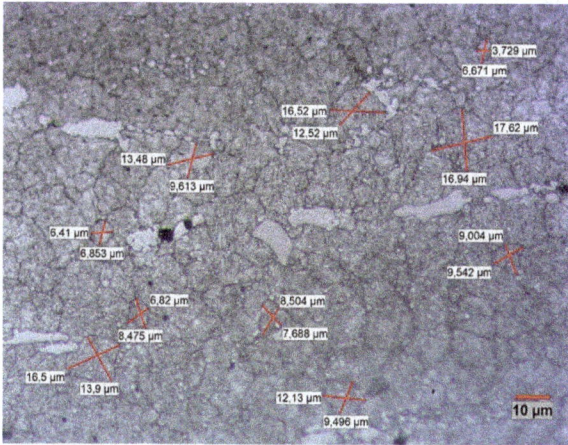

(b)

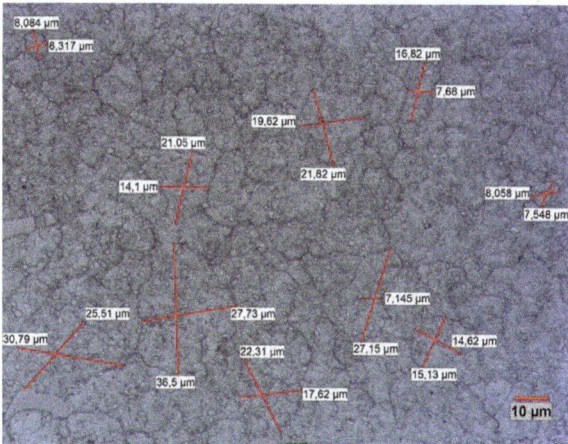

Figure 7: Smaller grains in the rim zone (a) and coarser grains in the core (b) of the MM 1.3344 solid bar leading to higher TRS values in the rim zone and lower values in the core.

Comparing the different materials, it is obvious that the PM 1.3344 exhibits considerably higher TRS values than its melting metallurgical counterpart MM 1.3344 across all measuring positions and orientations. In some cases, the values are even by a factor of 2 - 3 higher. Only at position AR the forging fiber reduces the difference between the two materials, although the PM 1.3344 values here still achieve 33% more in TRS. Compared with the highly ductile 1.2344 tool steel, PM 1.3344 shows comparable TRS values. This points out, that PM 1.3344 combines high toughness with high strength and hardness together.

Conclusions and Outlook

In the present work, a PM 1.3344 solid bar, a PM 1.3344/1.2344 composite bar and a melting metallurgically produced 1.3344 solid bar were compared regarding the hardness and transverse

Materials Research Forum LLC
https://doi.org/10.21741/9781644902837-14

rupture strength after heat treatment. The experiments point out that the PM 1.3344 has an exceptionally favorable combination of hardness, strength and toughness, coming from its fine-grained microstructure with a high amount of small and homogeneous distributed wear-resistant carbides, which underlines its suitability as working rolls in cold rolling mills or as wear parts in twin screw extruders, respectively pellet mills. This microstructure permits uniform quenching and tempering over the entire powder layer, which, in addition to a high hardness, results in high transverse rupture strengths, partly higher by a factor of 2 - 3 compared to its melting metallurgical counterpart MM 1.3344. Thereby the PM 1.3344 results depend neither on the orientation nor on the depth of the samples.

The composite PM 1.334/1.2344 and the solid bar PM 1.3344 hardly show any differences in their TRS values. This indicates that the heat treatment for the composite bar should be optimized to further improve the TRS of the core zone. In a first step the tempering temperature could be increased. This is expected to decrease the hardness and to increase the transverse rupture strength of 1.2344. Moreover torsion tests should be performed to better meet the alternating loads in the specific application. It is expected that the composite bar can withstand significantly higher torsional forces than the PM solid bar, resulting in considerable service life extensions. This fact has been proven in twin screw extruders for years. Regardless of the results of further investigations, at today's raw material prices a composite part made of expensive PM powders and cheaper conventional steels offers cost savings without compromising on quality.

References

[1] C. B. Boyer, Historical Review of HIP equipment, in: Hot Isostatic Pressing - Theory and applications, Proceedings of the 3rd International Conference on Hot Isostatic Pressing, Osaka, 1991, pp. 465-510. https://doi.org/10.1007/978-94-011-2900-8_68

[2] The Evolution of HIP - Commemorating the First Hot and Cold Isostatic Pressing Vessels, The American Society of Mechanical Engineers, 1985

[3] Introduction to Hot Isostatic Pressing technology - A guide for designers and engineers, European Powder Metallurgy Association, 1st Edition, 2014

[4] F. X. Zimmerman, HIP Equipment for Industrial Applications, in: High-Pressure Science and Technology. Springer, Boston, MA, 1979, pp. 1711–1717. https://doi.org/10.1007/978-1-4684-7470-1_209

[5] B. Hofer, A. Altay, M. Hamentgen, A. Schütz, Hot isostatic pressing and its applications, in: Proceedings of the 4th ESTAD, Düsseldorf, 2019

[6] H. Winkler, Einfluss der Pulverfraktion auf die mechanischen Eigenschaften von pulvermetallurgisch hergestelltem Schnellarbeitsstahl, Diploma thesis, Montanuniversität Leoben, 2007

[7] B. Hofer, F. Bigolin, M. Hamentgen, H. Stremming, C. Zühlke, Measuring the gas content in HIP components and impurities in the argon- and chemical reacted gas used to compacting near netshape parts and castings. Presentation of the measurement technique, in: Proceedings of the 11th International Conference on Hot Isostatic Pressing, Stockholm, 2014

[8] DIN EN ISO 3327:2009: Hardmetals – Determination of transverse rupture strength, Beuth Verlag, Berlin, 2009

Materials Research Forum LLC
https://doi.org/10.21741/9781644902837-15

Fabrication of Large Three-Dimensional Flow Path Structure Using SS Flexible Tube

Ryunosuke Kitamori[1,a*], Mitsuo Okuwaki[1,b], Shigeki Tsuruoka[1,c], Takuya Nagahama[1,d]

Metal Technology Company, Ltd., Tamamura-machi, Sawa-gun, Gunma-ken, Japan

[a]rkitamori@kinzoku.co.jp, [b]okuwaki@kinzoku.co.jp, [c]s_tsuruoka@kinzoku.co.jp, [d]tnagahama@kinzoku.co.jp

Keywords: Design, Design Flexibility, Flexible Tube, Cooling Channel, Corrosion, Corrosion Resistance, Flow, Flow Rate, Stainless Steel (SS), Bend, Bendable, Heat Exchange, Aluminum, Diffusion Bonding

Abstract. Aluminum is a commonly used material when cooling channels are required due to its high thermal conductivity and light weight. However, aluminum has a low corrosion resistance which is a key factor to be aware of when a cooling system utilizing water is required. Metal Technology Co. Ltd. (MTC) currently manufactures a variety of aluminum products in which stainless steel (SS) tubes are embedded and bonded with HIP to improve corrosion resistance while maintaining a light weight. As the demand for complex three-dimensional (3D) cooling channels increases, the complexity creates limitations on creating the cooling channels using the current methods and processes. 3D flow paths can be fabricated if SS tubes are bent and then welded together, but the assembly process for HIP is complex. Complex, thin-walled SS flexible tubes are not able to withstand the pressures applied during diffusion bonding. MTC, has successfully developed an alternative for making internal 3D cooling channels with SS flexible tubes.

Introduction

As the demand for smaller more powerful chips in the semiconductor market grows, so does the need for more complex cooling systems for the equipment that makes these chips. Some of the requirements for these new systems are low weight, high thermal conductivity and efficiency, and high corrosion resistance. With that in mind, flow path design in recent years is changing from a simple structure to a 3D structure. This requires more sophisticated manufacturing techniques and technology that can deliver the desired results.

Aluminum is often used as the material of choice for temperature control parts because of its excellent thermal conductivity and light weight. However, corrosion resistance is an obstacle when aluminum is used. Therefore, SS tubing is embedded in the aluminum and integrated with HIP to produce a temperature control component that is light weight and has excellent thermal conductivity while improving corrosion resistance.

However, there are some problems in fabricating parts with 3D flow paths. First, the more complicated the SS tube becomes, the more difficult it becomes to bend. Further, SS tube must have a wall thickness that can withstand bending without failure. As SS is inferior to aluminum in heat transfer, the performance of the cooling component decreases as the thickness of the tube becomes thicker. With that in mind, we propose manufacturing a large 3D flow channel structure using SS flexible tubes to solve these problems.

Challenge

Deformation of SS flexible tube

When manufacturing a 3D flow channel structure using SS flexible tubes, it is necessary for thin-walled SS flexible tubes to withstand the high temperature and pressure environment found during HIP treatment. We carried out various tests to determine the diameter and wall-thickness of the SS flexible tube to determine the parameters for temperature and pressure during HIP.

Heat exchange efficiency of SS flexible tubes and SS tubes
Temperature control parts are required to have high performance regarding heat exchange. It is anticipated that a cooling water channel with a SS flexible tubes wider surface area, in comparison with a standard SS tube, will provide better cooling performance. To confirm this, samples were produced to confirm the cooling performance and flow when using SS flexible tubes.

Experiment
Deformation of SS flexible tube
The HIP conditions were as follows: temperature at 600°C (1112°F) and pressure at 30Mpa. The aluminum "flowed" into the gaps in the SS flexible tube at all diameters. Figure 1 shows the HIP processing results.

Outline of test
· Flexible tube shape: Diameter of 16mm (0.63"), 29.5mm (1.16"), 54mm (2.13").
· Base metal: A1050 stress-relieving material is used.
· HIP-condition: Main thermal temperature 600°C (1112°F), pressure 30MPa

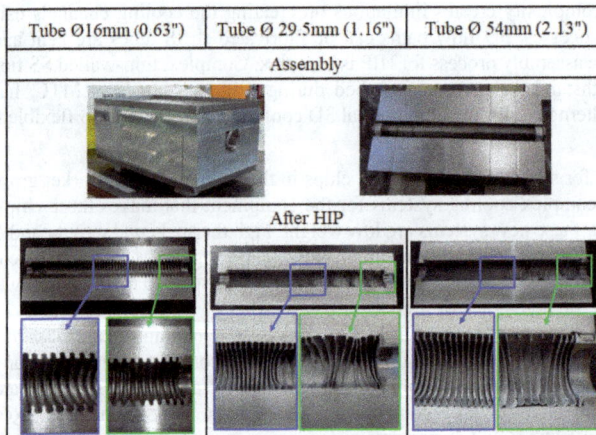

Tube Ø16mm (0.63")	Tube Ø 29.5mm (1.16")	Tube Ø 54mm (2.13")
Assembly		
After HIP		

Figure 1. SS Flexible Tube Diameter Difference HIP Test

As shown in Fig. 1 the **54mm (2.13")** tube after HIP shows significant deformation and was unable to obtain the desired HIP results. The 29.5mm (1.16") tube did obtain the desired HIP results, but also showed some deformation. The small diameter tested at 16mm (0.63") was able to obtain the desired HIP results while maintaining the original shape of the flexible tube.

Heat exchange efficiency of SS flexible tube and SS pipe
(1) Confirmation of cooling performance using a small sample
Considering the above-mentioned diameter differences during HIP testing, a small sample was manufactured using a flexible tube with an outer diameter of 8.5mm. The SS tube is a thin

thickness tube that can be bent and has a shape similar to a flexible tube with an outer diameter of 8.5mm. Figure 2 shows the shape of each small sample.

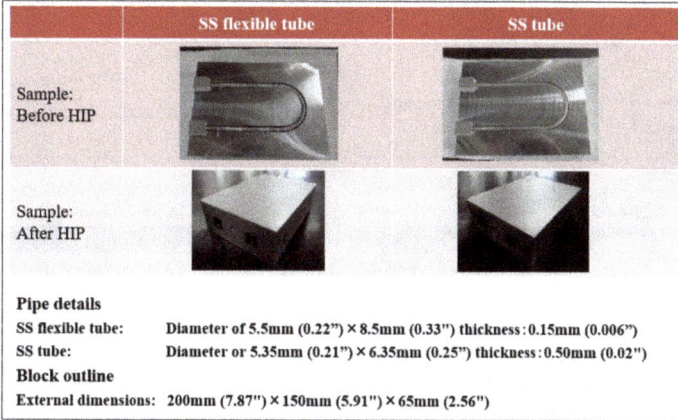

	SS flexible tube	SS tube
Sample: Before HIP		
Sample: After HIP		

Pipe details
SS flexible tube: Diameter of 5.5mm (0.22") × 8.5mm (0.33") thickness : 0.15mm (0.006")
SS tube: Diameter or 5.35mm (0.21") × 6.35mm (0.25") thickness : 0.50mm (0.02")
Block outline
External dimensions: 200mm (7.87") × 150mm (5.91") × 65mm (2.56")

Figure 2. Small Sample Geometry

To confirm the state of heat exchange efficiency, the samples were heated to measure the temperature change while cooling water flowed through the tubes. Heated by a hot plate, the cooling water flowed at 4L/min after the small sample reached 100°C (212°F). The temperature was measured with a thermocouple. Figure 3 shows the measurement positions for the temperature check and the schematic diagram of the test. Figure 4 shows the temperature measurement results.

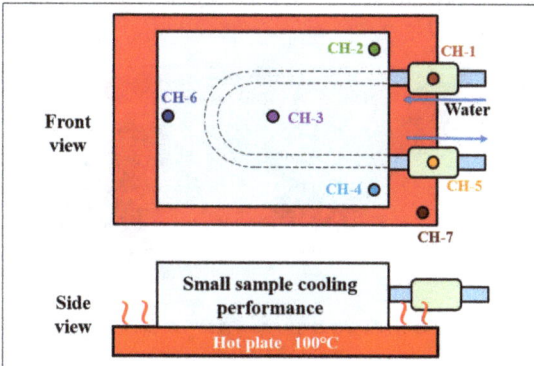

Figure 3. Test Overview

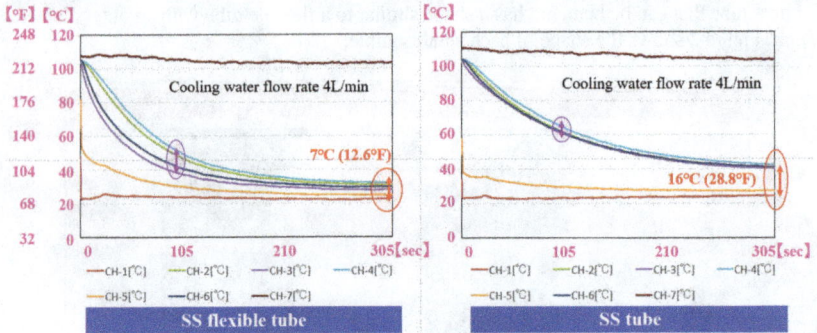

Figure 4. Actual Temperature Results of SS Flexible Tube and SS Tube

As cooling water flowed through the samples, a comparison of temperature at 105 seconds was done. While the standard SS tube temperature was between 60°C~65°C (140~149°F), the SS flexible tube was between 38~48°C (100.4~118.4°F). Although some variation was evident, this data shows that the SS flexible tube was more than 10°C (18°F) cooler at this point.

When a temperature comparison of the samples was done at 305 seconds, the temperature variation in the standard SS tube was about 16°C (28.8°F), while that of SS flexible tube was about 7°C (12.6°F). The results show a difference of 9°C (16.2°F) in the cooling capacity difference at the measured interval.

(2) Water flow vector analysis of SS flexible tube

Analysis was used to investigate the effects of unevenness on the surface of the flexible tube on the water flow. The analysis model was the SS flexible tube used in (1) "Confirmation of cooling performance using a small sample". Figure 5 shows the analysis model and analysis conditions. Analysis results are shown in Fig. 6.

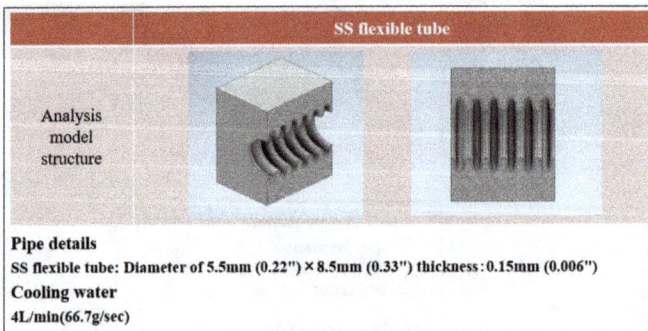

Figure 5. Analysis Model and Conditions

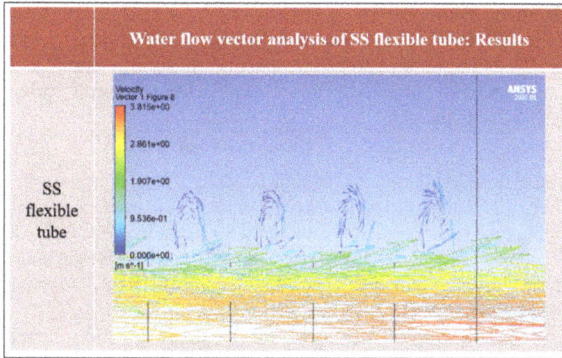

Figure 6. Water Flow Vector Analysis

Figure 6 shows that the water flow near the surface of the tube forms a vortex in the opposite direction to the flow due to the uneven design of the flexible SS tube. In addition, the flow velocity slows where the vortices are generated.

It was found that the flexible SS tube has higher cooling performance than the standard SS tube in the test results of the small samples described above, but it is believed that better cooling performance may be further improved by adjusting the water flow.

Adaptation to Large Parts Using SS Flexible Tube

Larger sample parts were tested using the processing conditions and test results described above.

Figure 7. Photographs are shown on the product.
- External dimensions: 400mm (15.75")×220mm (8.66")×30mm (1.18")
  Total length after assembly: 1200mm (47.2")
  Cooling tubes were placed on the curved surface of R2000 (78.7").
- Manufacturing process: Pre-processing → Assembly → HIP → De-capsulation →Machining → Wire cutting.

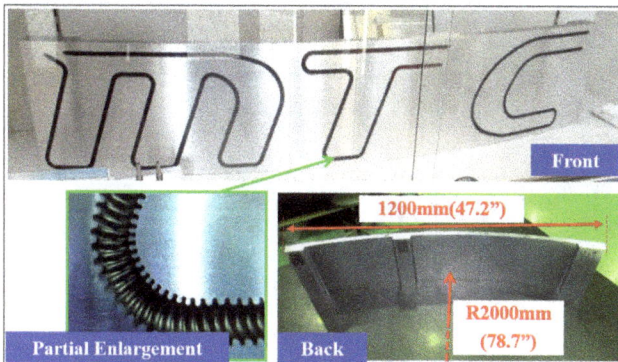

Figure 7. Large Part Sample Using SS Flexible Tube

Materials Research Forum LLC
https://doi.org/10.21741/9781644902837-15

It was determined that the aluminum flowed into the gaps of the SS flexible tube without crushing the tube and the desired HIP effect was obtained in spite of the large parts. Samples were designed so that the letters "M T C" were exposed on the curved surface during cutting. It has been determined that manufacturing large 3D flow channel structures with complicated shapes is possible by using SS flexible tubes.

Conclusion

It was proven that the flow of aluminum, without crushing the uneven portion of the SS flexible tube is viable when diameters of 16mm (0.63") or less are used. Regarding heat exchange efficiency, SS flexible tubes were superior to standard SS tubes from the test results obtained from the small samples and a cooling capacity difference of 9°C (16.2°F) was observed under the same conditions. Finally, it was determined that the manufacture of large 3D flow channel structures using SS flexible tube is possible.

Moving forward, the challenges of increasing the diameter and flow characteristics will be investigated. Once these challenges have been met, the range of products benefitting from this technology will expand and many more benefits will be achieved.

Acknowledgment

We would like to thank Mr. Nagasawa and Mr. Hashimoto for providing the experimental data.

Hot Isostatic Pressing - HIP'22

Materials Research Proceedings 38 (2023) 107-112

Materials Research Forum LLC

https://doi.org/10.21741/9781644902837-16

Powder Metallurgy HIP and Extrusion Study of FeCrAl Alloy for Accident Tolerant Fuel Cladding

Shenyan Huang[1,a*], Evan Dolley[1,b], Steve Buresh[1,c], Ian Spinelli[1,d], Jason Leszczewicz[1,e], Marija Drobnjak[1,f], Mike Knussman[1,g], Raul B. Rebak[1,h]

[1]GE Research, 1 Research Circle, Niskayuna NY, 12309, USA

[a]huangs@ge.com, [b]dolley@ge.com, [c]buresh@ge.com, [d]spinelli@ge.com, [e]Jason.Leszczewicz@ge.com, [f]marija.drobnjak@ge.com, [g]knussman@ge.com, [h]rebak@ge.com

Keywords: FeCrAl, Accident Tolerant Fuel, Nuclear Materials, Powder Metallurgy, HIP

Abstract. Powder metallurgy HIP and extrusion processing conditions were investigated in a FeCrAl alloy (Fe-Cr-Al-Mo based PM-C26M) for accident tolerant fuel cladding applications. The powder size range, HIP temperature ($900 \sim 1100°C$), HIP time (2 hours vs. 4 hours) were varied in lab-scale experiments, using argon gas atomized powder. The resulting density, grain size distribution, and retained plastic strain were characterized to recommend the HIP condition with full consolidation, fine grain size for formability, and good powder economics. The subsequent extrusion process was also studied in the temperature range of $950 \sim 1050°C$ as well as post annealing conditions. Based on the microstructure evaluation, sensitivity of HIP and extrusion process parameters were better understood.

Introduction

During the last decade, the international nuclear materials community was studying the possibility of replacing the zirconium based nuclear fuel cladding with materials that would be more tolerant of a loss-of-coolant accident. One of the most promising alternatives is to use tubes of FeCrAl alloys for the cladding of the fuel [1-3]. FeCrAl alloys have extraordinary resistance to attack by superheated steam and offer good mechanical and corrosion resistance properties under light water reactors normal operation conditions [4,5]. FeCrAl alloys were studied extensively in the 1960s for nuclear materials applications [2]. Even then it was recognized that the traditional alloy fabrication process of melting, casting, and forging was not the most promising for this family of alloys. The powder metallurgy process was more efficient since it resulted in reliable quality extruded bar stock that can be later pilgered into 5 m long 10 mm diameter tubes with a wall thickness down to 0.3 mm [4]. The objective of this work was to understand the processing window and sensitivity of HIP and extrusion parameters for the newer PM-C26M FeCrAl alloy, which does not have oxide dispersion strengthening nano particles that are characteristics of the more traditional APMT alloy.

Influence of HIP temperature, time, powder size range

Argon gas atomized PM-C26M powder was obtained from Sandvik with the actual composition of Fe-12.2Cr-5.7Al-2Mo in weight percent. As-received powder had a size range of $20 \sim 500$ μm and tap density of 4.56 g/cm^3 (or relative tap density 64%). D10, D50, D90 of as-received powder were 44.6, 103.0, 206.6 μm, respectively. A representative powder morphology and cross-section microstructure is shown in Fig. 1. Powder particles are mostly spherical with some satellites and trapped round argon pores inside powder, revealing equiaxed grain structure regardless of individual powder particle size. Two powder size ranges were studied for HIP processing, including the as-received $20 \sim 500$ μm and the $20 \sim 250$ μm (coarse particles sieved out). Subscale cylindrical HIP canisters of 19 mm diameter and 51 mm height were filled with powder, outgassed, helium leak checked, and then HIP'ed to get full powder densification. Detailed parameters are summarized in Table 1. All of the HIP cycles were performed with a simultaneous linear ramp of temperature (10°C/min) and pressure followed by a furnace cool after the HIP cycle in the Moly 5.5 inch Moly HIP unit at American Isostatic Presses. Two canisters with different powder size ranges and the same HIP parameters were HIP processed in the same cycle, to prevent process variation or introduction of a lurking variable. Ten HIP cycles were performed for the study.

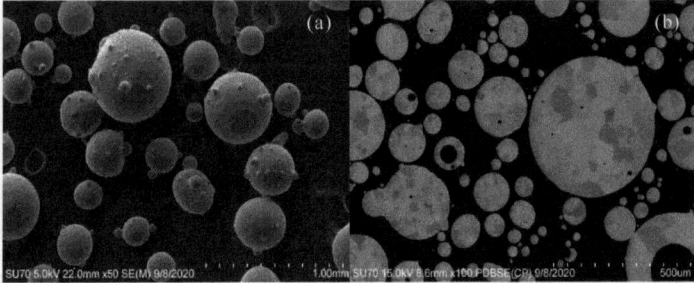

Figure 1: (a) Secondary electron image of as-received PM-C26M argon gas atomized powder particles; (b) backscattered electron image of powder cross sections.

Table 1: Detailed powder size range and HIP parameters in the HIP study. The average grain size measured by EBSD is specified for select conditions.

	900°C 15ksi 2hr	900°C 15ksi 4hr	950°C 15ksi 2hr	950°C 15ksi 4hr	980°C 15ksi 2hr	980°C 15ksi 4hr	1050°C 15ksi 2hr	1050°C 15ksi 4hr	1100°C 15ksi 2hr	1100°C 15ksi 4hr
20~250μm	X	X	--	--	X	X, 20μm	X	X, 21μm	X	X, 26μm
20~500μm	X	X	X	X, 18μm	X	X, 20μm	X	X, 28μm	X	X, 35μm

After HIP, the canisters were machined to remove canister material. The obtained cylinders of the consolidated PM-C26M were measured for density using the Archimedes method. Fig. 2(a) plots the resulting densities as a function of HIP temperature, time, and powder size range. It clearly shows that HIP temperature higher than 980°C achieved full density. HIP time of 2 hours seems sufficient for the lab-scale small HIP cans, as longer HIP time of 4 hours didn't result in a density increase. A similar trend is observed for the two powder size ranges studied. A minimum of 3 hours HIP time may be required for larger HIP cans, considering additional time needed for equilibrium temperature/pressure. Lower HIP temperatures of 900°C and 950°C failed to achieve full densification, even at 4 hours of HIP time. An example of 900°C HIP microstructure is displayed in Fig. 2(b), indicating residual porosity inside the prior powder particles or at the powder particle boundaries. Porosity inside the prior powder particles are likely the trapped argon gas porosities originated from the gas atomization process that were not fully closed by the HIP cycles due to insufficient temperature for the diffusional process.

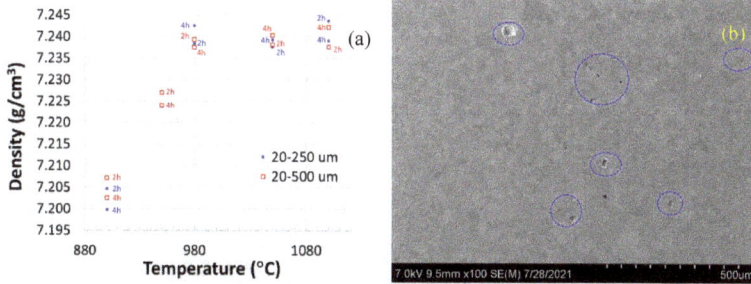

Figure 2: (a) Post-HIP density of PM-C26M as a function of HIP temperature, time, powder size range; (b) secondary electron image of the resulting microstructure with 900°C/15ksi/4hr HIP condition and 20~250 μm powder size range. Blue circles highlight the residual porosity.

The resulting grain structures as a function of HIP temperature and powder size range are presented in Fig. 3, with HIP temperature from high to low and two powder size ranges side by side for the same HIP condition. It is observed that the 20 ~ 500 μm powder size range with higher fraction of coarse powder particles generates severe grain growth in the larger powder particles (>200 μm) at higher HIP temperatures of 1050 and 1100°C. As a result, some coarse powder particles even became single grain, since there were few secondary precipitates in PM-CM26 to effectively pin grain boundary migration and thermal energy at 1050 and 1100°C HIP temperatures dominated grain growth. In comparison, the level of grain growth in the 20 ~ 250 μm powder size range was less severe due to lack of large powder particles above 250 μm. These large grains or single-grain powder particles could add difficulty to the processability of the subsequent bar extrusion. When HIP temperature was reduced to 980°C, finer grain structure absent of single grain powder particles was observed, which appeared to be adequate as-HIP grain structure for the subsequent extrusion. 980°C HIP didn't reveal a large sensitivity with respect to the powder size range. Although 900°C HIP shows an even finer grain size, the residual porosities from partial consolidation could cause crack initiation in extrusion and thus was not further considered.

Detailed EBSD characterization was performed on the select conditions to quantify the grain size distribution, average grain size, and evaluate the retained plastic strain. Hitachi SU-70 FEGSEM with Oxford Aztec Symmetry EBSD was used. A 20kV accelerating volage, 0.25 μm step size, 1000 x 750 μm scan area were applied. Additional maps were acquired for 1050 and 1100°C HIP with a 20 ~ 500 μm powder size range to ensure good statistical representation of the coarse grains. The total grain count for each condition was in the range of 2000 ~ 5000. Average grain size was quantified by line intercept method and the calculated mean intercept length is specified in Table 1, which verified the fine grain size (20 μm) in 980°C HIP and the grain growth (up to 35 μm average grain size) in the coarse powder size at higher HIP temperatures. The cumulative frequency of grain size plot and grain size distribution plots in Fig. 4 further indicate the grain size shifting to the large size and wider distribution with higher HIP temperature and coarser powder. Kernel average misorientation maps were also acquired to qualitatively reveal the retained plastic strain (not shown herein due to space limitations). In general, the retained plastic strain for the characterized conditions was quite low. Grain boundary maps for HIP temperatures at or above 980°C suggest ~20% of low angle boundaries (3 ~ 10°) and Σ3 special boundaries in the HIP conditions and largely recrystallized structure. While the 950°C HIP condition shows ~25% low angle boundaries localized at the space between powder particles.

The measured average Vickers microhardness for all the studied conditions was in the range of 211 ~ 221 HV. There was no obvious hardness difference in various conditions given the standard deviation. One additional drawback of having coarse powder particles was found to be the coarse powder particles transporting to the upper portion of HIP can during vibrational powder packing step before outgassing step. Coarse powder particle separation is a concern for large HIP parts. Therefore, the most reasonable condition found in the study seems to be 20 ~ 250 µm powder size range and 980°C/15ksi/4hr HIP condition.

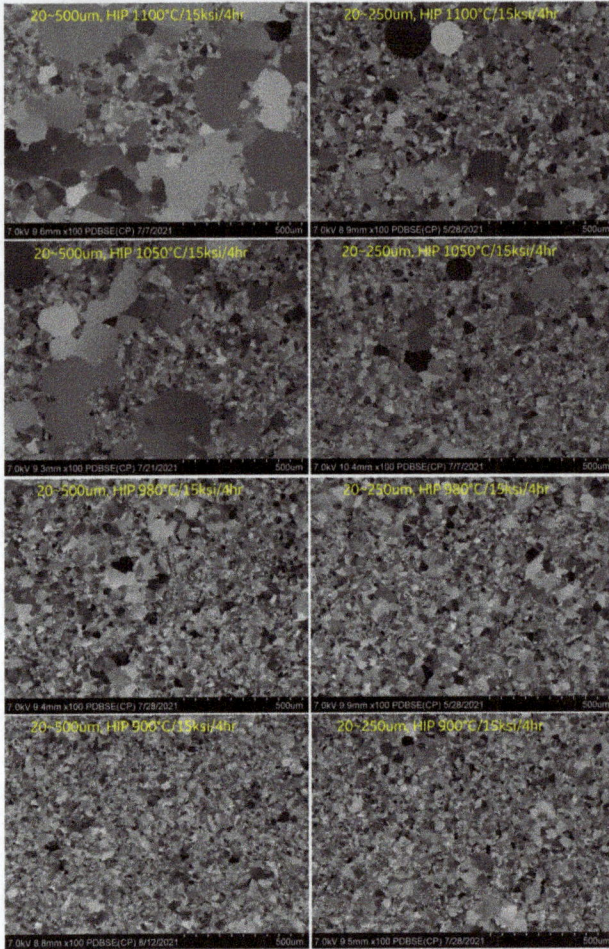

Figure 3: Backscattered electron images of as-HIP grain structures as a function of powder size range and HIP parameters.

Materials Research Forum LLC
https://doi.org/10.21741/9781644902837-16

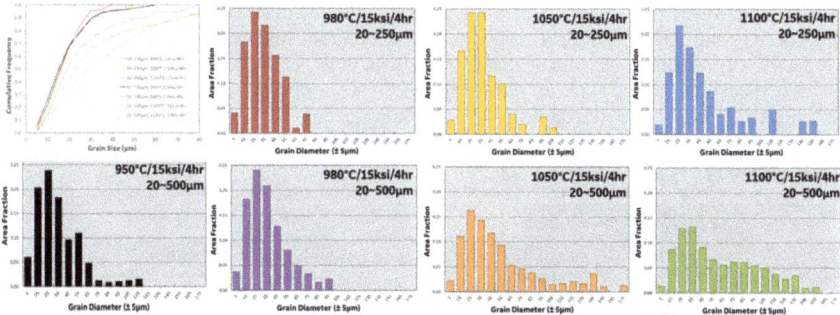

Figure 4: EBSD measured cumulative frequency of grain size and grain size distribution plots for 7 select conditions.

Influence of Extrusion temperature and Post Annealing

Another PM-C26M powder lot atomized by Oerlikon was used for the subscale HIP and extrusion study. Powder composition (Fe-11.9Cr-6.0Al-1.9Mo) was within the specification range. A narrower powder size range (up to 106 μm) was used. Three cylindrical HIP cans containing ~3 kg powder were HIP'ed at 980°C/25ksi/4hr. The as-HIP grain structure (Fig. 5a) seems much finer possibly caused by finer powder size compared to the coarse powder cut discussed earlier (Fig. 3). After machining to remove the canister material, the three HIP billets were hot extruded at 950, 1000, 1050°C, respectively. Hot extrusion was performed at GE Research on a 1250 Ton Horizontal Loewy Extrusion Press. Starting billet sizes ranged from 3.25" to 3.31" diameter (due to HIPing variation). A nominal 3.5" diameter container was utilized in the press along with a ~1.13" diameter conical die. After extrusion the bars were air cooled to room temperature. Figure 6 shows the as HIP'd powder canisters and the extruded bars (after grit blast).

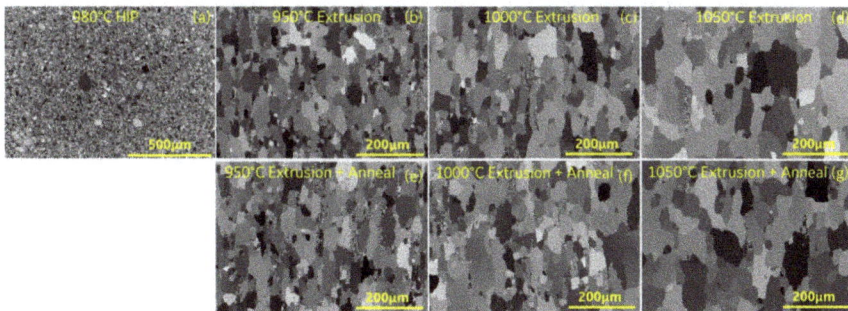

Figure 5: Backscattered electron images of grain structures in (a) the 980°C HIP billets, (b) 950°C extrusion, (c) 1000°C extrusion, (d) 1050°C extrusion; (e) 950°C extrusion followed by 800°C/30min annealing, (f) 1000°C extrusion followed by 800°C/30min annealing, (g) 1050°C extrusion followed by 800°C/30min annealing.

Post extrusion annealing was performed at 800°C for 30 minutes based on prior experience. The as-extruded and post-annealed microstructures are shown in Fig. 5. In general, the extruded microstructure largely shows low retained plastic strain and most grains appeared coarser than asHIP condition. Dynamic recrystallization and grain growth apparently have occurred during the extrusion process. There is still small fraction of fine grain pockets/chains remaining in the extruded microstructure. The fraction of fine grain pockets decreases with an increase in extrusion temperature from 950°C to 1050°C. The post-annealed grain structures looked similar to the as extruded grain structures, suggesting that 800°C/30min didn't promote further grain growth at the fine grain pockets. The grain size resulting from 950C extrusion seems adequate for the subsequent tube pilgering process. More detailed future work is required to optimize extrusion conditions, including lower extrusion temperatures, other post anneal conditions, and sensitivity of powder size distribution.

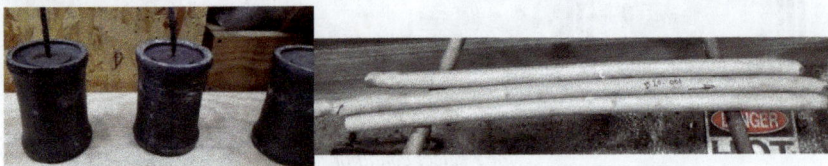

Figure 6: Three HIP billets and resultant extruded bars produced at different extrusion temperatures.

Summary

HIP and extrusion process parameter window was studied for FeCrAl alloy PM-C26M, a candidate material for accident tolerant fuel cladding. Consolidated microstructure proved sensitive to HIP temperature and powder size range, while HIP times of 2 hours vs. 4 hours had little effect. The most reasonable HIP condition that achieved full density and fine grain size was 20 ~ 250 μm powder size range and 980°C/15ksi/4hr HIP condition. HIP temperature below 950°C failed to produce full densified material, while severe grain growth was a concern at HIP temperature above 1050°C particularly for coarse powder range 20 ~ 500 μm. 950°C extrusion yielded reasonable grain structure for the subsequent tube pilgering process. Future work is recommended to optimize extrusion conditions, powder size distribution, mechanical properties of extrusion, and the sensitivity of ductility for the subsequent pilgering process.

Acknowledgement

This work was funded by US Department of Energy, National Nuclear Security Administration, under award numbers DE-NE0008823 and DE-NE0009047. Sandvik and Oerlikon are acknowledged for producing argon gas atomized C26M powder utilized in this work. American Isostatic Presses is acknowledged for HIP cycle processing.

References

[1] K. A. Terrani, Accident tolerant fuel cladding development: Promise, status, and challenges, Journal of Nuclear Materials, Vol. 501, (2018) 13-30.
https://doi.org/10.1016/j.jnucmat.2017.12.043

[2] K. G. Field and S. A. Briggs (2020), "Radiation Effects in FeCrAl Alloys for Nuclear Power Applications," Elsevier Comprehensive Nuclear Materials 2nd edition, Vol. 3.
https://doi.org/10.1016/B978-0-12-803581-8.11613-3

[3] R. B. Rebak, Accident Tolerant Materials for LWR Fuels, Elsevier, 2020.

[4] S. Huang, E. Dolley, K. An, D. Yu, C. Crawford, M. A. Othon, I. Spinelli, M. P. Knussman, and R.B. Rebak (2022), "Microstructure and tensile behavior of powder metallurgy FeCrAl accident tolerant fuel cladding," Journal of Nuclear Materials 560 (2022) 153524.
https://doi.org/10.1016/j.jnucmat.2022.153524

[5] R. B. Rebak, T. B. Jurewicz, M. Larsen, L. Yin, Zinc water chemistry reduces dissolution of FeCrAl for nuclear fuel cladding, Corrosion Science, Vol. 198 (2022) 110156.
https://doi.org/10.1016/j.corsci.2022.110156

Materials Research Forum LLC
https://doi.org/10.21741/9781644902837-17

Preparation of High-Quality Mo-Nb (Ti/Ni-Ti) Sputtering Target by Hot Isostatic Pressing

Zhanfang Wu[1,a], Lida Che[1,b*], Jing He[1,c], Haofeng Li[1,d],
Pengjie Zhang[1,e], Xiangyang Li[2,f]

[1]CISRI HIPEX, Haidian District, Beijing, P. R. China

[2]CISRI, Haidian District, Beijing, P. R. China

[a]wuzhanfang@hipex.cn, [b]chelida@hipex.cn, [c]hejing@hipex.cn, [d]lihaofeng@hipex.cn,
[e]zhangpengjie@hipex.cn, [f]lixy@cisri.com.cn

Keywords: Hot Isostatic Pressing; High Quality Sputtering Target; Powder Metallurgy; Density; Grain Size

Abstract. Mo-Nb, Mo-Ti, Mo-Ni-Ti sputtering target samples were prepared using Mo powder, Nb powder, Ni powder and Ti powder as raw materials by hot isostatic pressing (near net forming). The density, phase composition, microstructure and element distribution of Mo-Nb, Mo-Ti, Mo-Ni-Ti sputtered target samples were measured by optical microscope, scanning electron microscope and energy dispersive spectrometer. The experimental results show that hot isostatic pressing can prepare Mo-Nb target at lower temperature, and the obtained samples have finer grains, higher density and more uniform element distribution than that of non-pressure sintering. High-quality Mo-Nb, Mo-Ti, Mo-Ni-Ti sputtering targets (density>99%, grain size<100 μm) were obtained by hot isostatic pressing.

Introduction

With the rapid development of electronic information and related industries, the market demand for high-quality sputtering targets is increasing. Sputtering targets are widely used in many fields, such as plane display backplane, semiconductor, integrated circuit, recording medium and energy battery[1-4]. With the continuous expansion of application fields, the added value of sputtered targets is constantly improving. Therefore, the requirements for its performance and technical indicators are more stringent. The refractory metal Mo sputtering target has the advantages of high melting point, high elastic modulus, low thermal expansion coefficient, low resistivity and good thermal stability. It is widely used in the sputtering coating industry and occupies an important position. It is an indispensable raw material for products such as flat panel displays, semiconductors (gate electrodes, connecting wiring, diffusion barrier layers), microelectronics, solar cells, etc., and has broad market prospects[5]. In the electronics and information industries, such as semiconductors, integrated circuits and liquid crystal displays, which require high target quality, have strict requirements on the purity, density, grain size and ingredient uniformity of the target[6]. The films sputtered from pure Mo targets have some problems in terms of corrosion resistance (discoloration) and adhesion (film peeling). A large number of studies show that adding Nb, Ti, Ni and other metals to molybdenum can make its specific impedance, stress, corrosion resistance and other properties balanced[7-9].

There are many disadvantages in the production of Mo alloy targets by the traditional sintering method, such as high target porosity, low yield, coarse grains, and poor organizational uniformity, which makes it difficult to guarantee product quality. Hot isostatic pressing (HIP) is a kind of technology that takes inert gas as the pressure medium, places the product in a closed container, and applies equal pressure to the product under the joint action of high temperature (~2000°C) and high pressure (~200MPa) to sinter the product [10]. In this study, Mo-Nb, Mo-Ti and Mo-Ni-Ti alloy targets will be prepared by hot isostatic pressing, the effects of preparation methods on the microstructure, density and element distribution of Mo-Nb, Mo-Ti and Mo-Ni-Ti alloy targets will be studied, so as to provide process references for the preparation of high-quality Mo alloy targets.

Test materials and methods

Test materials

The average powder particle diameter of Mo powder selected in this test is 3.3 μm~5.0 μm, Nb powder is 45 μm~75 μm, Ni powder is 15 μm~45 μm and Ti powder is 45 μm~75 μm. The specific components of the powder are shown in Table 1, and the purity of all powders is above 99.95%. The powder morphologies of Mo powder, Nb powder, Ni powder and Ti powder are shown in Fig. 1. It can be seen from Fig. 1 that under the observation of the scanning electron microscope, the morphology of Mo powder is relatively uniform, and the particles are spherical or nearly spherical. The morphology of the Nb powder particles is relatively non-uniform in size, and most of them are irregular blocks or prismatic structures. The Ni powder particles are agglomerated, which requires reduction heat treatment before mechanical mixing. The morphology of the Ti powder particles is similar to that of the Nb powder particles, and the size is relatively non-uniform and irregular prismatic structure.

Table 1. Impurity element content of test metal powder / ($\times 10^{-6}$)

Powder	C	O	N	H	Fe	Si	Cu	Al	Ca	Mg	S
Mo	17	120	5	12	29	7	1	3	2	2	/
Nb	105	750	108	11	36	15	4	11	/	/	11
Ni	50	610	150	10	50	5	1	1	/	/	/
Ti	110	960	120	15	40	3	/	/	/	10	/

Figure 1. Powder morphologies by SEM
Mo powder; (b) Nb powder; (c) Ni powder; (d)Ti powder

Materials Research Forum LLC
https://doi.org/10.21741/9781644902837-17

Sample preparation method and process

The first step is mixing the powder of Mo+Nb, Mo+Ti, Mo+Ti+Ni according to the specified weight ratio. Then load the mixed powder into a vertical V-type mixer, where the three-dimensional mixing is carried out under the protection of argon. The stirring power is 1.1 KW, the stirring speed is 30 rpm~40 rpm, and the three-dimensional mixing time is 4 h. The oxygen in the air sucked by the metal powder shall be strictly controlled during the whole mixing process. The mixed Mo-Nb powder is divided into two groups, one group is used for non-pressure sintering, the other group is used for hot isostatic pressing. The specific preparation process of all samples is shown in Table 2. The sintering temperature of Mo-Nb sample is 1600 °C~1800 °C, and sintering in reducing atmosphere for 6~8 h. The hot isostatic pressing process of Mo-Nb sample is to put the mixed powder into a soft steel capsule, preheat the powder at 400 °C~500 °C, and conduct degassing sealing welding to make the vacuum degree less than 6×10^{-3} Pa. Put the capsule (after degassing sealing welding) into the furnace for hot isostatic pressing. The hot isostatic pressing pressure is 120 ~180 MPa, temperature is 1150~1300 °C, and holding time is 2~4 h. The hot isostatic pressing process of Mo-Ti sample is as follows: the hot isostatic pressing temperature is 950 ~1100 °C, pressure is 120 ~180 MPa, and holding time is 2~4h. The hot isostatic pressing temperature for Mo-Ni-Ti sample is 900~1050°C, the pressure is 120 ~180 MPa, and holding time is 2"4 h. After the preparation of the samples, the metallographic and density tests are carried out on all samples. The metallographic samples are polished with 400, 600, 800, 1000 and1200 mesh water abrasive paper, then cleaned with alcohol, dried and finish polish. The microstructure observation and energy spectrum analysis of the samples are carried out with optical microscope and Zeiss SUPER55 scanning electron microscope.

Table 2. Preparation Process of Samples

Sample	Process	Process parameters
Mo-Nb	Non-pressure sintering	Sintering temperature: 1600°C-1800°C, sintering in reducing atmosphere: 6~8h.
	HIP	Soft steel for capsule with a thickness of 3mm, hot isostatic pressing temperature is 1150°C~1300°C, pressure is 120MPa~180MPa, holding time is 2~4h.
Mo-Ti	HIP	Capsule thickness: 3mm, hot isostatic pressing temperature is 950°C~1100°C, pressure is 120MPa~180 MPa, holding time is 2~4h.
Mo-Ni-Ti	HIP	Capsule thickness: 3mm, hot isostatic pressing temperature is 900°C~1050°C, pressure is 120MPa~180 MPa, holding time is 2~4h.

Results and discussion

Density and oxygen content

Figure 2 shows the density test results of the samples. It can be seen from the figure that the density of Mo-Nb sample obtained by non-pressure sintering is only 96.5%, which does not meet the requirement ($\geq$99%) for high-quality sputtering targets. The density of the Mo-Nb sample prepared by hot isostatic pressing is 99.8%. The density of the Mo-Ti obtained by hot isostatic pressing is 99.8%, which is close to the theoretical density of the sample. The density of Mo-Ni-Ti obtained by hot isostatic pressing is 99.1%, which is relatively low, but it has also reached the density requirement for high quality sputtering target. By analyzing and comparing the densification degree of these samples, it is shown that hot isostatic pressing can greatly reduce the porosity and increase the density. Table 3 shows the detection values of oxygen content of the samples. It can be seen from the table that the oxygen content of the sample prepared by non-pressure sintering reaches 1380 ppm, exceeding the required range. The reason may be related to the fact that the samples were not protected by reducing atmosphere in time during high temperature sintering for a long time.

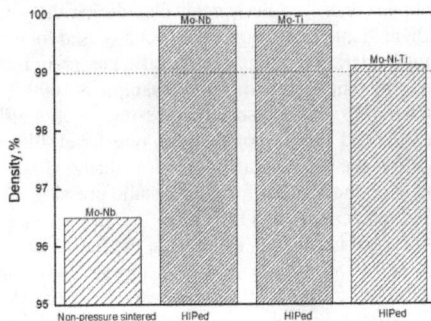

Figure 2. Results of sample density

Table 3. Oxygen content of samples

Sample	Oxygen content $/(\times 10^{-6})$
Mo-Nb（non-pressure sintered）	1380
Mo-Nb（HIPed）	790
Mo-Ti（HIPed）	860
Mo-Ni-Ti（HIPed）	760

Microstructure

Figure 3 shows the metallographic micrographs of Mo-Nb, Mo-Ti and Mo-Ni-Ti samples. It can be seen from Fig. 3 (a) that the grain size distribution of Mo-Nb sample prepared by non-pressure sintering is wide, the larger grain size reaches 200 μm, and the smaller grain size is less than 50 μm. Fig. 3 (b) shows the microstructure of the Mo-Nb sample prepared by hot isostatic pressing. It can be seen from the figure that the grain size is small and evenly distributed, and the grain size is 20 μm~100 μm, the original size of the powder is basically maintained. Mo-Ti and Mo-Ni-Ti samples prepared by the hot isostatic pressing process have uniform microstructure and no obvious pores and oxide inclusions, which are shown in Fig. 3 (c) and Fig. 3 (d). Compared with these metallographic micrographs, it can be found that the samples prepared by non-pressure sintering have more pores and larger grain size. In Fig. 3 (a), many particles and pores (black part) are distributed along the grain boundary. Referring to the measurement results of the oxygen content of the sample in Table 3, it can be preliminarily determined that the oxide in the sintered body is the main reason for the low density. In order to find the reason for the low density of non-pressure sintering, the internal factors of the low density of Mo-Nb sample prepared by non-pressure sintering process will be determined by scanning electron microscope and energy spectroscopic analysis.

Materials Research Forum LLC
https://doi.org/10.21741/9781644902837-17

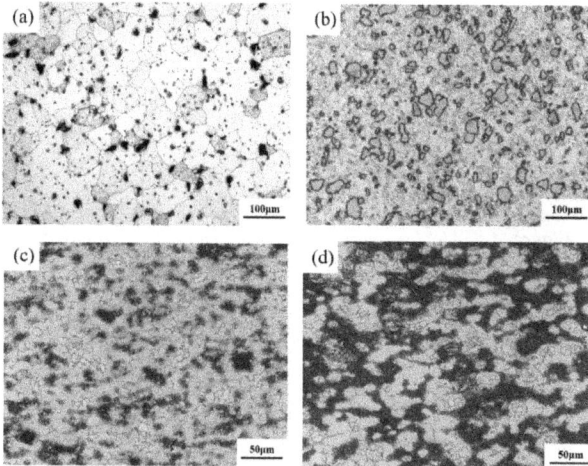

Figure 3. Metallographic micrographs of samples (a) Mo-Nb (non-pressure sintered); (b) Mo-Nb (HIPed); (c) Mo-Ti (HIPed); (d) Mo-Ni-Ti (HIPed)

Distribution of chemical elements

Figure 4 shows the SEM analysis results of Mo-Nb sample prepared by non-pressure sintering. It can be seen from Fig. 4 that there is obvious aggregation of Nb elements in the Mo-Nb alloy sample. From the energy spectrum analysis results in the figure, it shows that Nb elements aggregate at the grain boundaries in the form of NbO_x. The fine niobium powder is easy to absorb oxygen, and it is difficult to control the oxygen content. Therefore, these oxides may be brought into the powder particles or generated during the sintering process. According to relevant literature reports[11], the deoxidation of niobium oxide in the sintering process is divided into two stages: the first stage has only a small amount of weight loss, and the second stage starts at 1995 K, a large amount of weight loss occurs in this stage. The formation of NbO is better than that of NbO_2 under high temperature and low oxygen conditions. If the high temperature hydrogen reduction sintering method is used, due to the moisture contained in the hydrogen, the free Nb is easily combined with it to form NbO_2 during the heating process. Therefore, this explains the phenomenon that a large amount of NbO_x accumulates at the grain boundary during the sintering process of Mo-Nb to reduce the density. Figure 5 shows the element distribution of the Mo-Nb sample prepared by hot isostatic pressing. It can be seen that the element distribution of the Mo-Nb sample prepared by the hot isostatic pressing process is uniform. Figure 6 is the scanning electron microscope analysis of the fractures of Mo-Ti and Mo-Ni-Ti samples. It can be seen that the fracture of Mo-Ti sample is flush and there is no obvious pore. While the fracture of Mo-Ni- and there is an obvious hole embedded inside, which is about 10 μm in length and 2~3 μm in width. The reason for the formation of voids is probably due to the reaction between Mo and Ni to form Mo-Ni brittle phase at high temperature during hot isostatic pressing, which is also the reason for the relatively low density of Mo-Ni-Ti. Therefore, the research results show that the hot isostatic pressing parameters for Mo-Ni-Ti need to be optimized.

Materials Research Forum LLC

https://doi.org/10.21741/9781644902837-17

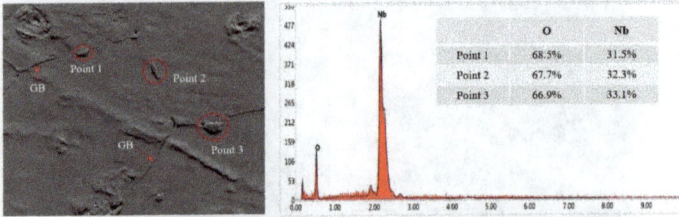

Figure. 4 SEM-EDS analysis of Mo-Nb (non-pressure sintered)

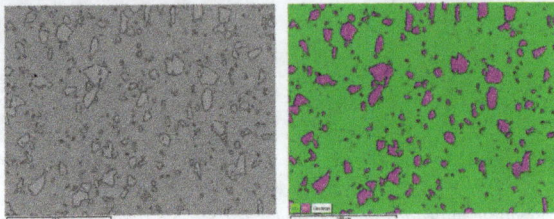

Figure. 5 SEM-EDS analysis Mo-Nb (HIPed)

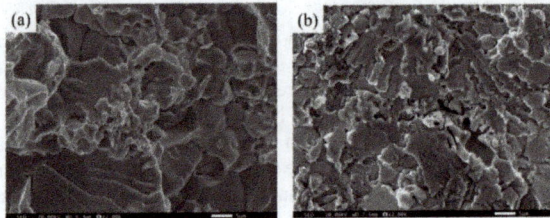

Figure. 6 SEM analysis of sample fracture
(a) Mo-Ti (HIPed); (b) Mo-Ni-Ti (HIPed)

There are some studies on the influence of microstructure and density on the sputtering performance of Mo target. Huang H S *et al.* [12] studied the effect of the preparation process on the microstructure and sputtering performance of the Mo target by using several preparation methods. The research results show that hot isostatic pressing can prepare Mo targets with a density close to the theoretical density. In contrast, through hydrogen sintering, the density of the Mo target reaches only 95%. In addition, due to the lower hot isostatic pressing temperature, grain growth is not obvious after densification. Therefore, the microstructure of the Mo target processed by HIP is very fine, and the average grain size is below 100 μm. The study of sputtering performance shows that the Mo target produced by hot isostatic pressing has the highest deposition rate, followed by the hydrogen reduction sintered Mo target, and the asrolled Mo has the lowest deposition rate. Researchers have carried out relevant research on the microstructure and sputtering performance of Mo target, but there are few detailed reports on the preparation method, microstructure analysis and phase analysis. In particular, the rapid development of hot isostatic pressing technology in recent years has opened up a new way to prepare Mo and most refractory metal targets by using this technology to produce high-quality targets.

Materials Research Forum LLC
https://doi.org/10.21741/9781644902837-17

Conclusion

In this study, the effects of preparation methods on the microstructure, density and element distribution of Mo-Nb, Mo-Ti and Mo-Ni-Ti targets are analyzed by means of optical microscope, scanning electron microscope and energy spectrometer. The conclusions are as follows:

(i) Hot isostatic pressing can prepare Mo-Nb target with finer grain, higher density and more uniform element distribution than non-pressure sintering at lower temperature.

(ii) A large number of niobium oxide particles gather at the grain boundary of MoNb alloy prepared by non-pressure sintering, which is the main reason for the insufficient density. An important source of niobium oxide is the moisture in hydrogen during sintering.

(iii) High-quality Mo-Nb, Mo-Ti and Mo-Ni-Ti sputtering targets with density >99% and grain size <100μm can be prepared by hot isostatic pressing.

Acknowledgements

The authors thank CISRI HIPEX for providing the hot isostatic pressing equipment. This work was supported by China Iron & Steel Research Institute Group, and Project No. 2020018.

References

[1] Z.Q. Chu. The present status and development trend of magnetron sputtering target, J. Metal Materials and Metallurgy Engineering, 39(2011) 44-49.

[2] Z.Y. Shang, X. Jiang, Y.J. Li. Integrated circuit manufacturing sputtering target, J. Rare Metals, 29(2005) 475-477.

[3] H. Wang, M. Xia, Y. Li, X. Liu, X. Cai. Application and preparation technology of refractory metal sputtering target, China Molybdenum Industry, 34(2019) 64-67.

[4] [4] B.H. Zhao, H.B. Fan, J Sun. TFT-LCD source for producing molybdenum thin film sputtering and target, J. China Molybdenum Industry,35 (2011) 7-11.

[5] X.M. Dang, G. An, J. Li. Effect of niobium powder granularity and sintering method on Mo-Nb alloy, J. Powder Metallurgy Industry, 34(2016) 271-276.

[6] Y. P. Ning, W. Wang, Y. Sun. Effects of substrates, film thickness and temperature on thermal emittance of Mo/substrate deposited by magnetron sputtering, J. Vacuum.128 (2016)73-79. https://doi.org/10.1016/j.vacuum.2016.03.008

[7] F. Yang, K.S. Wang, P. Hu. research status and development trend of high-purity molybdenum sputtering targets, J. Thermal Processing, 39(2013) 10-12.

[8] G. An, J. Sun, R. Liu, J. Li, W. Cao. Study on microstructure and properties of Mo thin films sputtered by two kind of Mo targets, J. Material Sciences, 8(2018) 709-717. https://doi.org/10.12677/MS.2018.86084

[9] [9] H.L. Liu, M. Bai, G.T. Sun. Basic performance research and reapplication development of molybdenum target after magnetron sputtering, J. China Resources Comprehensive Utilization, 10(2021) 73-79.

[10] B. Williams, F.L. Han. Recent trends in Hot Isostatic Pressing (HIP): Processing and application, J. Powder Metallurgy Industry, 32(2014) 464-468.

[11] T. K. Roy, A. Awasthi, N. Krishnamurthy. Studies on sacrificial and carbon deoxidation of niobium, J.International Journal of Refractory Metals & Hard Materials 004(22): 251-256. https://doi.org/10.1016/j.ijrmhm.2004.08.002

[12] H.S. Huang, H. C. Tung, C. Chao. Correlating fabrication processes to the microstructure and deposition properties of Mo sputtering targets, J. China Steel Technical Report, 26(2013): 59-66.

Hot Isostatic Pressing - HIP'22
Materials Research Proceedings 38 (2023) 120-130

Materials Research Forum LLC
https://doi.org/10.21741/9781644902837-18

Simulation-Based Manufacturing of Near-Net-Shape Components and Prediction of the Microstructural Evolution during Hot Isostatic Pressing

Yuanbin Deng[1,a] *, Anke Kaletsch[1,b] and Christoph Broeckmann[1,c]

[1] RWTH Aachen University, Institute for Materials Applications in Mechanical Engineering, Augustinerbach 4, 52062, Aachen, Germany

[a]y.deng@iwm.rwth-aachen.de, [b]a.kaletsch@iwm.rwth-aachen.de, [c]c.broeckmann@iwm.rwth-aachen.de

Keywords: Hot Isostatic Pressing (HIP), Discrete-Element-Method (DEM), Capsule Filling, Finite-Element-Method (FEM), Densification, Phase Transformation/precipitation, Modelling, Simulation

Abstract. Following the development of hot isostatic pressing (HIP) with integrated rapid cooling technology, it is now possible to combine consolidation of encapsulated powder and subsequent heat treatment in a single step. In this study, the influences of pressure and cooling rate on the microstructural evolution of martensitic and duplex steels during the entire HIP process with rapid cooling are investigated. Besides the microscopic investigation of the microstructure, a material model for finite element (FE-) simulations was developed to numerically correlate the understanding based on experiments and to predict the final shape of the HIP component. This FE-simulation was additionally employed in the capsule design to achieve net-shape production of the components with complex geometries. The agreement between the experimental and simulated results validated the method to be able to ensure a near-net-shape product and to monitor microstructural development during HIP and rapid cooling.

Introduction

Hot isostatic pressing (HIP) is widely applied in powder metallurgy (PM) to produce components with high requirements on the final geometry and properties, as it densifies metal powder to a fully dense bulk component with a homogeneous, isotropic, and fine-grained microstructure [1]. The Powder HIP involves filling of metal powder into a metallic capsule and consolidating of the powder to full density in a HIP unit. Due to the limited filling density of the powder after pre-densification, the capsule shrinks up to about 30% in volume during the HIP process. Thereafter, the capsule has either reached its net-shape geometry or has to be further processed as a semi-finished product. Conventional capsules for powder HIP are welded from sheet metal and thus are restricted to produce components with simple geometries. With the development of additive manufacturing (AM), it is possible to build predestined complex-shaped HIP capsules, thus, the net-shape parts can also be achieved by using optimized capsule geometry with the help of numerical simulation [2]. Besides near-net-shape manufacturing, the adjustment of the microstructure of the final component in order to achieve desired mechanical properties remains challenging. Based on the increasing attraction of applying HIP with integrated rapid cooling to produce materials like martensitic and duplex stainless steels, where specific microstructural constituents significantly determine the properties of the component, a further development of the macroscopic HIP models is required to predict the evolution of the microstructure. To realize this, the understanding of the kinetics of the studied microstructural constituents as well as the heat transfer between the cooling gas and the component in the HIP unit are indispensable. The former provides sufficient parameters that can be implemented in the simulation model. The latter offers

Materials Research Forum LLC
https://doi.org/10.21741/9781644902837-18

an accurate description of the temperature distribution in the component. As a result, the evolution and distribution of phases within the whole component can be accurately predicted during cooling. This work focuses on the simulation-based manufacturing of near-net-shape components with the prediction of the microstructural evolution during hot isostatic pressing. Capsule filling, near-net-shape manufacturing and the prediction of microstructural constituents has been studied by an extruder screw as an example for a typical component. The additively produced capsule of AISI A11 was filled with AISI L6 powder. The phase transformation during the cooling stage was modelled and numerical studied. By employing a second demonstration component that is made of AISI 318LN, filled in a conventionally produced capsule, a precipitation model was developed, which allows the prediction of σ-phase volume fraction with the consideration of heat transfer analysis during HIP with integrated rapid cooling. The numerical methods developed here are expected to support powder HIP of near-net-shape components with complex geometries and desired microstructure by adjusting the capsule design.

Materials and modelling

Materials, capsule geometries and HIP cycles
In this work, the capsules with two different geometries (Fig. 1 (a) and (b)) were hot isostatically pressed using the integrated rapid cooling option. These capsules (a) and (b) were made of two different kinds of steels and filled with two different powders, respectively, as listed in Table 1. The HIP cycles for both capsules had a heating rate of 500 K/h, a holding temperature of 1150°C, and a pressure of 100 MPa. The holding times for capsules/powder AISI A11/AISI L6 and AISI 304/AISI 318LN were 3 h and 2 h, respectively. To determine the model parameters and to manufacture near-net-shape components, several HIP-units were used in this work, which are listed in Table 2. All capsules were cooled down rapidly in the HIP vessel subsequent to the holding stage.

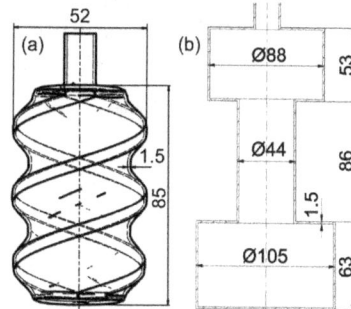

Figure 1: Capsule geometries

Table 1: Applied materials of capsules (a) and (b) during HIP.

Capsule geometry	Capsule	Powder/Bulk
(a)	Martensitic tool steel AISI A11 (FeCrV10)	Martensitic tool steel AISI L6 (DIN 56NiCrMoV7)
(b)	Stainless steel AISI 304 (DIN X5CrNi18-10)	Duplex stainless steel AISI 318LN (DIN X2CrNiMoN22-5-3))

Table 2: HIP units and types.

Capsule	HIP unit, type	Integrated rapid cooling unit
(a)	QIH-15L Quintus Technologies, Sweden	√
(b) and cycles for the determination of model parameters	QIH-9, Quintus Technologies, Sweden	√
	Shirp 20/30-200-1500, ABRA Fluid AG, Switzerland and reconstructed by Cremer Thermoprozessanlagen GmbH	√

Materials Research Forum LLC
https://doi.org/10.21741/9781644902837-18

Modelling of particle flow and capsule filling

Since the powder filling process can be seen as a granular system and each particle in this system moves independently, it is difficult to predict the behavior of the granular system using continuum mechanical models. In this context, the discrete approach developed for numerical modelling of granular materials at particle scale, which is generally referred to as discrete element method (DEM) [3, 4], has become a powerful and reliable tool. DEM is a particle method based on Newton's laws of motion. In this method, particles are defined as soft particles that can move with six degrees of freedom (three for translational and three for rotational movements), but only undergo elastic deformation. In this way, the simulation of the contact between particles considers a cohesive force (F_{ij}^{Coh} in Eq. (1)), which is calculated by a contact model and a cohesion model. The Hertz–Mindlin model, as a widely used contact model, describes the normal interaction by a nonlinear relationship which represents the elastic contact behavior between particles. When the distance d_{ij} between two particles is smaller than their contact distance $R_i + R_j$, the model is used to calculate the Hertz–Mindlin contact force F_{ij}^{HM}. The simplified model of Johnson-Kendall-Roberts (SJKR) as the cohesion model adds an additional normal force contribution to the Hertz-Mindlin model. If two particles are in contact, the SJKR model adds an additional normal force F_{ij}^{Coh} which helps to maintain the contact. In addition to the contact forces, gravitational acceleration is also considered and the gravitational force F_i^{Grav} is calculated [5]. In this way, all forces on a particle can be calculated as follows:

$$F_{ij}^{sum} = F_{ij}^{HM} + F_{ij}^{Coh} + F_i^{Grav} \qquad (1)$$

Newton's equations of motion are used to determine the new positions of the particles. The equations of the interaction forces and moments in a time step for particle i are as follows

$$m_i \ddot{x}_i = F_i^{sum} \qquad (2)$$

$$I_i \dot{\omega}_i + \omega_i \times I_i \cdot \omega_i = t_i \qquad (3)$$

where m_i is the mass, $\ddot{x}_i$ is the acceleration, I_i is the Inertia tensor, ω_i is the angular velocity, $\dot{\omega}_i$ is the angular acceleration, and force F_i^{sum} and moment t_i are the sums of all forces and moments that act on the particle.

Modelling of densification during hot isostatic pressing

Many mechanisms, which take place during densification of powder in a HIP process such as particle rearrangement, plastic and viscoplastic deformations as well as diffusion, determine the final properties of the produced components. The contribution of each mechanism depends on the processing parameters in the individual HIP stage. Throughout the early stage, the particle rearrangement and the growth of necks at the contact points between the particles dominate the densification following the continuously increasing pressure and temperature. In the following stage, the necks grow until they impinge, and the individual pores are sealed off [6]. The rate-dependent deformation mechanisms overtake the granular dominated material behaviors at elevated temperature and pressure. Among these mechanisms, the two most significant ones that lead to the changes in density and volume are plastic yielding and viscoplastic deformation. Therefore, to precisely describe the material behaviors during HIP, the time independent plasticity model of Kuhn and Downey [7] was coupled with the time dependent plasticity (viscoplasticity) model of Abouaf et al. [8]. Both models have been proven by Wikman et al. [9] to be the best approaches to simulate the densification of porous media under the effect of high isostatic pressure.

Hot Isostatic Pressing - HIP'22
Materials Research Proceedings 38 (2023) 120-130

Materials Research Forum LLC
https://doi.org/10.21741/9781644902837-18

In previous publications of the authors [10–12], the constitutive equations and the implementation steps were given in detail.

Modelling of microstructural evolution during continuous cooling
The evolution of the microstructure, residual stresses and deformation during cooling is the outcome of a multi-physics process including different physical fields and the interactions between them. For a quantitative and predictive description in the cooling stage of the HIP cycle, a numerical approach is adopted, which considers the relevant themomechanical-metallurgical coupling. The featured interactions result in a time and position dependent stress/strain state. During quenching, thermal stresses arise due to the temperature gradient between the component's surface and core, which is determined by the heat transfer between the component and the pressurized gas. With the implementation of the experimentally determined heat transfer coefficient in the model, the inhomogeneous temperature distribution in the component that induces a heterogeneous microstructure after cooling is considered in the simulation.

The phase transformations have major contributions to the evolution of the microstructure as well as the geometrical distortion induced deformations and stresses during cooling. Hence, a reasonable description of the microstructural evolution that is important for the simulation of the cooling process requires a reliable modelling of the phase transformation kinetic. In this work, the diffusionless transformation (austenite→ martensite) is modelled by applying a slight modification to the Koistinen Marburger approach [13]. There is no need to formulate the model as a differential equation in this case because the transformation is time independent and thus, the equation (4) is directly implemented for the simulation of the transformation.

$$p_m = \tilde{p}_m \cdot \left[1 - exp\left\{ -\left(\frac{M_s(\lambda) - T}{b(\lambda)} \right)^{n(\lambda,T)} \right\} \right] \qquad (4)$$

In the equation (4), p_m is the volume fraction of martensite, $\tilde{p}_m$ is the maximum possible fraction of martensite which depends on the available austenite amount, M_s is the martensite start temperature, λ is the cooling parameter, T is the temperature ($T < M_s$), and finally b and n are material specific parameters. The new feature of this model adopted from [13] is that the effect of the cooling parameter on the kinetic of the martensitic transformation is considered. Unlike the martensitic transformation, the diffusion-controlled transformations (austenite →ferrite/perlite/bainite) are modeled using the modified Avrami equation as in [14]. The derivative form of the equation (5) is applied to adapt it to non-isothermal conditions by eliminating the time factor given in the original Avrami-model [14–16].

$$\dot{p} = n(T) \cdot \frac{\tilde{p}(T) - p}{\tau(T)} \cdot \left[ln \left(\frac{\tilde{p}(T)}{\tilde{p}(T) - p} \right)^{\frac{n(T)-1}{n(T)}} \right] \cdot f_1 \cdot f_2 \qquad (5)$$

In equation (5), $\dot{p}$ is the rate of the transformation and p is the volume fraction of the newly build phase. The kinetic parameters $n(T)$ and $\tau(T)$ depend on the nucleation and the growth rate. $\tilde{p}(T)$ is the maximum possible volume fraction of the product phase at the given temperature. This kind of formulation allows to consider the effect of different thermomechanical conditions on the kinetic of the transformation. For instance, the effect of the cooling rate on the transformation is modeled by a correction factor $f_1(\dot{T})$. Similarly, a factor f_2 could be applied to represent another interaction such as the effect of plastic deformations or the austenite grain size on the transformation kinetic.

Hot Isostatic Pressing - HIP'22   Materials Research Forum LLC
Materials Research Proceedings 38 (2023) 120-130   https://doi.org/10.21741/9781644902837-18

After minor modifications, this modified Avrami model can be further used to describe the phase precipitation during continuous cooling quantitatively [10], considering that the precipitation kinetics of the precipitate phase including nucleation and grain growth are controlled by the thermodynamic driving forces and the diffusion, respectively.

Numerical studies

Capsule filling

The DEM modelling technique was used to simulate the capsule filling. To consider the powder size distribution from the real powder, a simplified and upscaled powder model was created. As the number of particles in a DEM simulation cannot be defined arbitrarily due to the limitation of the computational power, it is not yet possible to simulate any given real system on a one-to-one particle level. To shorten the computational time, a coarse graining strategy was developed in the DEM modelling, which is capable of quantitatively describing static and dynamic behaviors of real powders. The size distribution of the modelled particles was adjusted to represent the experimental values obtained from real powder particles in different size categories (Table 3). To calibrate the scaling factors after grain-coarsening, the DEM model parameters (e.g., coefficient of restitution, friction coefficient, and cohesion energy density) of AISI L6 powder were adjusted in modelled experimental setups until the simulation results agreed with the experimentally measured data.

Table 3: Particle size distribution of real powder and modelled powder particles in four size categories

Particle size distribution of L6 powder					
Real powder		Simplified		Upscaled and modelled via DEM (Scaling factor 2.5)	
Radius [μm]	Weight [%]	Radius [μm]	Weight [%]	Radius [μm]	Weight [%]
0-57	11.80	54	15	135	15
58-99	32.50	84	30	210	30
100-149	38.70	114	35	285	35
150+	17.10	150	20	375	20

Simulation of the filling process with capsule geometry (a) was carried out in the LIGGGHTS software v3.8.0, which is a parallel particle simulator developed by DCS Computing GmbH, Austria. Figure 2 shows the different stages of the powder filling process in an extruder screw shaped capsule using a 3D geometry model. In this model, ~1.6 Mio. particles were filled into the capsule, while pre-consolidation processes such as vibration and tapping were not considered. As shown in Fig. 2 (a) and (b), the particles flowed a long distance from the inserted position to their final location. The fine powder particles rearranged after rebound and concentrated in the region close to the inner side of the screw thread. Particle segregation can be observed after the powder filling process. As the results of particle simulation, the relative density distribution in this extruder screw shaped capsule can be determined and then be mapped in the finite element model, which can be used to simulate the density evolution during the HIP process.

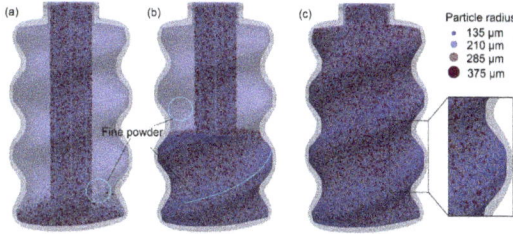

Figure 2: (a) initial, (b) intermediate, and (c) final stages of powder filling in the extruder screw shaped capsule.

Simulation of HIP including the prediction of phase fractions in final component

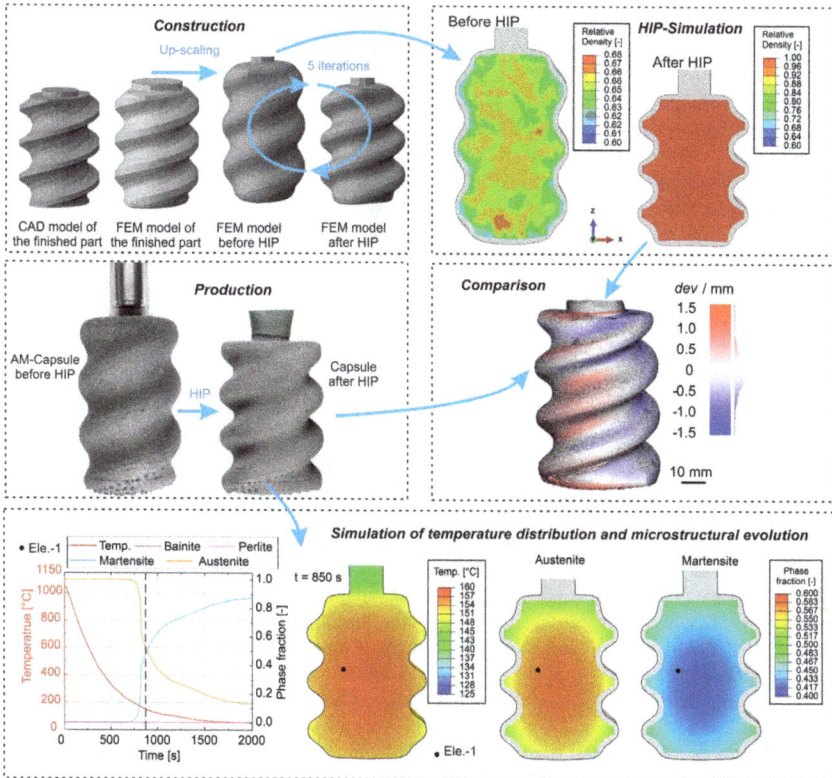

Figure 3: (a) evolution from CAD to optimized capsule geometry and near-net-shape geometry after HIP simulation; (b) relative densities determined in the simulation model prior to HIP (by DEM) and after HIP (by FEM); (c) screw extruder capsule before and after HIP; (d) comparison of optimized capsules for an extruder element in 'as HIP' state; (e) predicted temperature distribution and phase fraction distribution at 850 s.

Hot Isostatic Pressing - HIP'22   Materials Research Forum LLC
Materials Research Proceedings 38 (2023) 120-130   https://doi.org/10.21741/9781644902837-18

The HIP densification model based on FEM was implemented beforehand in the user- defined material model (UMAT) [17], which is used to simulate the material in ABAQUS 2020. The required material properties and model parameters were determined experimentally and reported in another publication of the authors [17]. To achieve a near-net-shape component, the modified capsule geometry was generated via "inverse optimization", which was introduced and validated in [12]. To save the computational time, a homogenous powder filling density of 0.62 from tapping tests was assumed for the capsule geometry optimization. Figure 3(a) shows the capsule design steps. As illustrated in Fig. 3(a), the target geometry was first generated based on the CAD-file of the extruder screw element. By adding the capsule region with a thickness of 1.5 mm, the FEM model of the finished part was then generated. Taking this geometry as the target geometry (iteration 1, before HIP), a deformed capsule geometry was obtained after the first simulation step and compared with target geometry. The capsule geometry was adjusted after each iteration and a final shape was predicted after each HIP simulation. After five iterations, the predicted component geometry (iteration 5, after HIP) is identical with the target geometry. The capsule geometry of the last step prior to HIP (iteration 5, before HIP) was used to produce the capsule by Electron Beam Powder Bed Fusion (PBF-EB) (Fig. 3 (c)) on an Arcam A2X machine with pre-heating at 850°C. The build job was conducted with a cross snake scan strategy and a layer thickness of 75 µm, while melting was performed with a line offset of 100 µm and an area energy of about 4 J/mm^2. The large freedom of product geometry as one of the advantages of Additive Manufacturing (AM) methods such as PBF-EB can be utilized if the geometry of the HIP capsule is designed using the aforementioned methods. Benefitting from the capsule design with simulation and production by PBF-EB, the final geometry of complex shaped component after HIP could be assured and the post-processing was expected to be eliminated.

The results shown in Fig. 3 proved the success of this method in achieving near-net-shape production. First, as described in section 3.1 the optimum size and shape for the capsule geometry was determined by FEM simulation. This geometry was used for the powder filling simulation by DEM. Fig. 3(b) shows the predicted powder density distribution after filling and the relative density distribution of the HIP'ed component obtained from the FEM model. From the DEM results, it can be seen that a region with high filling density is presented in the middle of the capsule, whereas the areas close to the inner wall of the screw thread show lower filling densities. The relative densities of the filling powder were between 0.62 and 0.68 with an average value of 0.64, which is 2% higher than the average filling density of 0.62 in a real extruder screw capsule. This 3D density database was imported into the simulation model to simulate the HIP process and obtain the final geometry of the component as well as the density distribution.

Secondly, a capsule of AISI A11 produced by PBF-EB was filled with the powder of AISI L6 and hot isostatically pressed experimentally. After rapid cooling, its final shape (Fig. 3(c)) was measured in 3D with an optical system to obtain the geometrical data file of the outer contour of the real HIP'ed part. By applying the software CloudCompare [18] the differences between the measured geometry and the simulated one were determined automatically and visualized in 3D, as shown in Fig. 3(d). The comparisons in all directions reveal a maximum average deviation of less than 2%.

The simulation of the HIP process was carried out not only using the HIP-model to predict the final geometry (section 2.3) but also the phase transformation model to calculate the microstructural evolution (section 2.4). After densification in the holding stage, the capsule was hardened by quenching directly from the HIP temperature to 100°C within 15 min. Most of the phase transformations that determine the final microstructure take place during the cooling stage, while the prerequisite of modeling phase transformation is the calculation of the spatial temperature distribution through the whole cooling process. As reported in a previous work of the authors [10], the experimentally measured heat transfer coefficients can be defined as thermal

Hot Isostatic Pressing - HIP'22

Materials Research Forum LLC

Materials Research Proceedings 38 (2023) 120-130

https://doi.org/10.21741/9781644902837-18

condition to simulate the temperature gradient on the HIP'ed components during cooling. By applying this model, the temperature distribution within the extruder screw was calculated, which allowed the prediction of the phase fraction distributions of bainite, pearlite, martensite, and austenite in the HIP'ed component. Figure 3(e) shows the simulated evolution of the microstructural constituents of a finite element in the center of the HIP'ed part during cooling as well as the temperature distribution within the component and the distribution of the volume fraction of austenite and bainite at a selected time (t=850 s). The final microstructure was predicted to be a mixture of martensite (~87.8%) and retained austenite (~12.2%), which agrees well with the data in the TTT diagram of AISI L6 reported by Eser [14]. This agreement confirms the accuracy of the phase transformation model in calculating the microstructural evolution.

To test the possibility of predicting the precipitate fraction by simulation using the developed phase transformation model, a capsule made of AISI 304 with the geometry (b) shown in Fig. 1 was filled by the powder of AISI 318LN and hot isostatically pressed [10]. Here the duplex stainless steel grade AISI 318LN was selected, as among all intermetallic phases σ–precipitates are the most critical to degrade the mechanical properties of duplex stainless steels. Similar to the previous case study, the densification of the powder, distortion of the capsule, as well as the temperature profile and the precipitation of the σ-phase inside of the being HIP'ed component were calculated in one step. A large temperature gradient up to 4 K/mm, as seen for instance in the temperature distribution at t=215 s in Fig. 4 (a), corresponds well to the distribution of σ–phase volume fraction after cooling (Fig. 4 (b)), despite a rather small predicted average amount of σ-phase of 11 ppm. Although only very limited amount of the σ–phase was precipitated in the simulation due to rapid cooling, its inhomogeneous distribution can still be clearly observed in Fig. 4 (b).

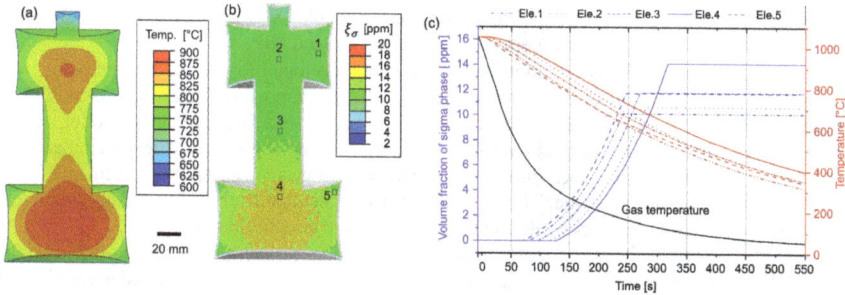

Figure 4: (a) predicted temperature distribution after 215 s; (b) predicted σ–phase distribution at ambient temperature; temperature and predicted σ–phase evolution profiles of five positions 1 -5 marked in (d) in the cooling stage [10].

To further understand the influence of the heat transfer, the predicted evolution of the σ–phase fractions of the five selected elements shown in Fig. 4(b) in the cooling stage were plotted in Fig. 4(c) and analyzed in detail. By observing their positions in the capsule shown in Fig. 4(b), it can be found that the determining factor of their cooling rates is not the shortest distances between them and their respective nearest capsule surface but their distances to the surface in the radial direction. Therefore, the most plausible explanation is that the heat transfer in the capsule depends predominately on the heat flux in the radial direction during rapid cooling.

This demonstrates the significance of a correct consideration of heat transfer in numerical models for the simulation of precipitation in complex shaped components. Moreover, a precisely described temperature distribution is the precondition to understand the anisotropic microstructure of near-net-shape powder HIP components.

Conclusions

In this work, the simulation with a coupled DEM and FEM model was used to calculate the powder density distribution after capsule filling in complex shaped capsules, to predict the capsule shrinkage during the HIP (i.e., the final geometry of the HIP'ed component), and to calculate the evolution of the microstructural constituents in the last rapid cooling stage. All simulation results show good agreements with the experimentally produced products. The main conclusions are summarized in the following.

- In DEM modelling, the particle size distribution was considered as an influencing factor in capsule filling, which leads to a clear visualization of the particle segregation in the filled capsule. The correct predicted powder density distribution will strongly improve the accuracy of the prediction of the capsule shrinkage during HIP simulation.

- Besides the powder density distribution, the optimization of the initial capsule shape is also an essential factor in achieving near-net-shape production of the HIP process. Based on FE-simulations, the capsule of an extruder screw was optimized in this study. Although the optimized capsule geometry is complex and requires precise production, capsules were still successfully produced by AM (PBF-EB), filled with powder and HIP'ed to receive final products.

- A maximum average deviation of less than 2 % was obtained by measuring and comparing the geometries of the simulated and experimentally produced components after HIP. This proves that the implementation of powder density distribution after capsule filling using DEM and capsule geometry optimization using FE-simulation enables near-net-shape production and at the same time combines the respective strengths of AM and HIP.

- In addition, a phase transformation model that was developed to predict the microstructural evolution was also added to the densification model to empower a macroscopic-continuum approach for the simulation of the powder HIP processes. The combined model, which considers different deformation mechanisms during densification and the precipitation or phase transformation kinetics during entire HIP cycle, was successfully implemented as a subroutine in the FE software. The accuracy of it was also confirmed by precisely predicting the shape of the final components and the phase volume distribution therein.

- Furthermore, the application of the simulated temperature distribution in the capsule with the consideration of experimentally determined temperature dependent heat transfer coefficients was proved to be able to further improve the accuracy of the prediction of phase volume fraction, especially for the HIP with integrated rapid cooling.

Acknowledgement

The authors would like to thank Ms. Berenice Kramer at LWT at Ruhr-University Bochum for HIP-cycles, Mr. Markus Mirz at IWM at RWTH Aachen University and Mrs. Marie Franke-Jurisch at Fraunhofer IFAM Dresden for the manufacturing of the extruder screw component. Our appreciation is also to Prof. Dr.-Ing Werner Theisen for valuable discussions.

Funding

This work was performed with the financial support from Germany Research Foundation (DFG) under the project No. 392860940. Thanks also go to the German Federal Ministry for Economic Affairs and Climate Action for financial support of the project IGF-21074 BG. Simulations were

Materials Research Forum LLC
https://doi.org/10.21741/9781644902837-18

performed with computing resources granted by RWTH Aachen University under project rwth1226.

References

[1] H. V. Atkinson and S. Davies, "Fundamental Aspects of Hot Isostatic Pressing: An Overview," *Metallurgical and Materials Transactions A*, vol. 31A, no. 12, 2000. https://doi.org/10.1007/s11661-000-0078-2

[2] E. Hernández-Nava, P. Mahoney, C. J. Smith, J. Donoghue, I. Todd, and S. Tammas-Williams, "Additive manufacturing titanium components with isotropic or graded properties by hybrid electron beam melting/hot isostatic pressing powder processing," *Scientific Reports*, vol. 9, no. 1, p. 4070, 2019. https://doi.org/10.1038/s41598-019-40722-3

[3] T. Pöschel and T. Schwager, *Computational granular dynamics: Models and algorithms,* 1st ed. Berlin, Heidelberg: Springer, 2010.

[4] P. A. Cundall and O. D. L. Strack, "A discrete numerical model for granular assemblies," *Géotechnique*, vol. 29, no. 1, pp. 47–65, 1979. https://doi.org/10.1680/geot.1979.29.1.47

[5] S. Luding, "Introduction to discrete element methods," *European Journal of Environmental and Civil Engineering*, vol. 12, 7-8, pp. 785–826, 2008. https://doi.org/10.1080/19648189.2008.9693050

[6] A. Bose and W. B. Eisen, *Hot consolidation of powders & particulates*. METAL POWDER INDUSTRIES FEDERATION, 2003.

[7] H. A. Kuhn and C. L. Downey, "DEFORMATION CHARACTERISTICS AND PLASTICITY THEORY OF SINTERED POWDER MATERIALS," *Int. J. Powder Metall.*, vol. 7, no. 1, p. 15, 1971.

[8] M. Abouaf, J. L. Chenot, G. Raisson, and P. Bauduin, "Finite element simulation of hot isostatic pressing of metal powders," *Int. J. Numer. Methods Eng.*, vol. 25, no. 1, pp. 191–212, 1988. https://doi.org/10.1002/nme.1620250116

[9] B. Wikman, A. Svoboda, and H. Å. Häggblad, "A combined material model for numerical simulation of hot isostatic pressing," *Comput. Meth. Appl. Mech. Eng.*, vol. 189, no. 3, pp. 901–913, 2000. https://doi.org/10.1016/S0045-7825(99)00406-5

[10] Y. Deng, J.-L. Zhang, A. Kaletsch, and C. Broeckmann, "Modelling and simulation of densification and σ-phase precipitation in PM duplex steel AISI 318LN during hot isostatic pressing," *Materials Today Communications*, vol. 29, p. 102901, 2021. https://doi.org/10.1016/j.mtcomm.2021.102901

[11] C. van Nguyen, Y. Deng, A. Bezold, and C. Broeckmann, "A combined model to simulate the powder densification and shape changes during hot isostatic pressing," *Computer Methods in Applied Mechanics and Engineering*, vol. 315, pp. 302–315, 2017. https://doi.org/10.1016/j.cma.2016.10.033

[12] S. Riehm *et al.*, "Tailor-made functional composite components using additive manufacturing and hot isostatic pressing," *Powder Metallurgy*, vol. 64, no. 4, pp. 295–307, 2021. https://doi.org/10.1080/00325899.2021.1901398

[13] A. Eser, A. Bezold, C. Broeckmann, I. Schruff, T. Greeb, "Tempering-Simulation of a thick-walled Workpiece made of X40CrMoV5-1 Steel," *Journal of Heat Treatment and Materials*, p. 127.

[14] A. Eser, *Skalenübergreifende Simulation des Anlassens von Werkzeugstahlen.* Dissertation, Aachen, 2014.

[15] M. Avrami, "Kinetics of Phase Change III. Granulation, Phase Change and Microstructure," *Journal of Chemical Physics*, p. 177, 1941. https://doi.org/10.1063/1.1750872

[16] A. Höfter, *Numerische Simulation des Härtens von Stahlbauteilen mit verschleißbeständigen Schichten.* Dissertation, Bochum, 2005. https://doi.org/10.3139/105.100294

[17] C. Broeckmann and T. Weißgärber, *Schlussbericht Maßgeschneiderte verschleißfeste Verbundbauteile durch Additive Manufacturing und heißisostatisches Pressen (Maß-HIP-3D): IGF Vorhaben Nr.: 21074 BG.* Aachen/Dresden, Germany, Herbst, 2022.

[18] D. Girardeau-Montaut, A. Maloney, and R. Janvier, *CloudCompare* (2019).

Hot Isostatic Pressing - HIP'22
Materials Research Proceedings 38 (2023) 131-140

Materials Research Forum LLC
https://doi.org/10.21741/9781644902837-19

Modelling of Powder Filling of HIP Canisters

Simon Chung[1,a*], Abheek Basu[1,b], Kieren Irvine[1,c], Peter Wypych[2,d],
David Hastie[2,e], Andrew Grima[2,f], Sam Moricca[1,g]

[1]Gravitas Technologies, Coniston, New South Wales, Australia

[2]University of Wollongong, Wollongong, New South Wales, Australia

[a]simon.chung@gravitas-tec.com, [b]abhi.basu@gravitas-tec.com,
[c]kieren.irvine@gravitas-tec.com, [d]wypych@uow.edu.au, [e]dhastie@uow.edu.au,
[f]agrima@uow.edu.au, [h]sam.moricca@gravitas-tec.com

Keywords: Hot Isostatic Pressing (HIP), Radioactive Waste, Discrete Element Method (DEM), Powder Filling, Idaho Calcine Simulant

Abstract. Hot-isostatic pressing (HIPing) of powders is achieved by placing them inside a canister, which is then evacuated and sealed. Canister filling is a critical step. Consistency in powder packing in the canister and increases in packing density will improve process efficiency and the predictability of HIP canister collapse, leading to less wastage. Understanding the effect of powder morphology, properties and characteristics on the can filling process and subsequent compaction is vital to optimizing canister design and the filling system. Conventionally, this has involved conducting numerous costly and time-consuming trial-and-error experiments. Computational modelling offers an alternative optimization path. Discrete element method (DEM) simulation of a powder filling process has been developed by GRI Inc. and its application to the US Department of Energy's radioactive Idaho calcine powders has been investigated. A comprehensive analysis of a non-radioactive simulant powder has been conducted. A DEM model was developed with validation of the model using experimental data obtained from the filling system development program.

Introduction

The HIP technology offers benefits for treating nuclear wastes and is the current baseline option for treating the Idaho calcines [1]. Approximately 4,400 m³ of highly radioactive Idaho calcine powders are stored in underground silos at the Idaho National Laboratory site [2]. The current plan is to extract the powders, load them into stainless steel canisters and consolidate them via HIPing to reduce the waste volume and form a more stable, dense wasteform suitable for transport and disposal.

Packing radioactive powder into HIP containers offers several challenges. Firstly, maximizing packing density is required to maximize throughput. The canister must be evenly packed so that shrinkage is predictable and controllable, especially radial shrinkage, as the HIPed canister will be placed into cylindrical metal overpack containers. Secondly, the filling system is a connect-disconnect process. It must be designed to control and minimize contamination from radioactive dust while operating remotely to minimize radiation exposure to the technical operators.

Conventionally, the development of bulk materials handling systems involves the construction of prototypes, followed by time-consuming trial-and-error experiments. Non-radioactive simulants of the Idaho calcine have been produced by the US Department of Energy (DOE) for process development trials. The simulant used in this research was created by a fluidized bed calcination process, similar to the Idaho radioactive waste calciner, to produce a powder with the same bulk material properties as the actual calcines. Hence, experimental data can be representative, and the test and design outcomes can be applied to an actual nuclear waste system.

However, the analysis of nuclear waste and the production of nuclear waste simulants is often expensive and made by specialized, small-scale production facilities. This makes it a scarce material for process validation and technology demonstrations. When a different stream of nuclear waste needs to be assessed, a slight change in the formulation may lead to completely different bulk material properties and behavior. Therefore, preliminary process developments are usually based on limited powder property data, such as particle size distribution and bulk density. Due to its scarcity, the waste simulant is often recycled numerous times to complete a testing program. Repetitive attrition can alter the flow properties of the recycled material, which may lead to results that are no longer representative of the actual nuclear waste. The trial-and-error approach has made process development for nuclear waste treatments challenging and sometimes prevented the development of new technologies.

Computational simulation based on the Discrete Element Method (DEM) offers an alternative approach. DEM is computationally demanding, which limited its industrial application in the past. However, with the advancement in computing hardware and software, DEM is increasingly being used as a tool for industrial process development. Previous studies have shown that a calibrated DEM model can accurately simulate industrial-scale applications [3-5]. This method has also been applied to scarce materials for research and development [6].

This study aims to investigate the potential of DEM for developing process systems that involve handling scarce and hazardous materials, a challenge commonly faced in the HIP industry. This paper presents a validated DEM model, which is employed to simulate an automated HIP can filling process.

Idaho Calcine Waste Simulant

Idaho calcine waste was converted from high-level aqueous radioactive wastes between December 1963 and May 2000 for volume reduction and safer temporary storage at the Idaho Nuclear Technology and Engineering Center (INTEC) [2]. A total of 4,400 m^3 of calcined solids were produced, retrieved and stored in Calcined Solids Storage Facilities (CSSFs). CSSFs are storage vaults consisting of stainless-steel storage bins within a concrete vault. There are seven CSSFs, with five reaching their total capacity. Although the calcined waste management strategy permits the interim waste storage of the Idaho calcine, the long-term plan is to retrieve the calcine, treat the calcine and ship the treated calcine to a long-term storage or disposal facility. Different calcine compositions across the bins reflect the composition of the fuel that was processed. In this study, alumina calcine (the predominant type and likely first to be extracted from the bins) waste simulant was selected as the sample material.

The extraction of the calcines from the CSSFs, their transfer to a treatment facility and loading into the HIP powder handling present challenges. Designing and implementing processes for these activities requires an understanding of the powder's bulk material behavior and chemical, physical and radiological characteristics. Properties of the radioactive calcine retrieved from the storage facility were published in a previous study [7]. Experiments were carried out in hot cells to determine the physical properties of the radioactive calcines, and the results were compared with non-radioactive calcine simulants. The published data includes bulk density, particle size distribution, instantaneous flow function, instantaneous wall friction angle and compressibility. This technical report concluded that there was little significant difference in properties between the simulated and radioactive materials. A more recent alumina calcine simulant has been acquired and characterized to provide certainty of data input to the DEM model.

Flow Property Testing and Characterization of Calcine Simulant

A simulant sample was tested and characterized using a Freeman FT4 powder rheometer (FT4) according to ASTM standards [8]. The waste simulant is an alumina-based material produced by a DOE contractor using a fluidized-bed calcination process similar to that used at INL. The

Hot Isostatic Pressing - HIP'22   Materials Research Forum LLC
Materials Research Proceedings 38 (2023) 131-140   https://doi.org/10.21741/9781644902837-19

composition is based on alumina calcine, minus the toxic elements, as given in Staiger and Swenson [2]. The simulant consists of free-flowing rounded, white, cream and orange-brown colored granules (Fig. 1) with low cohesiveness and excellent aerability. Particle size distribution determined by a Malvern Mastersizer 3000 revealed a bimodal distribution, with $D10 = 20.0$ µm, $D50 = 101$ µm, and $D90 = 363$ µm. Bulk density is found to be 1.21 g/cc. Compressibility and wall friction are shown in Fig. 5 and Fig. 6, respectively.

Figure 1: Microscopic image (×10 magnification) of Idaho calcine simulant

Additional experiments were conducted using a Flat Bottom Hopper Discharge Calibration Test Rig (FBH), previously validated to calibrate the DEM model for industrial applications [9]. The FBH was partitioned to have a footprint of 300×50 mm before 3.6 kg of simulant was loosely poured into the hopper. After the hopper was filled, the swing gate was opened to discharge the material, via a 15 mm width slot, into a base ring with an outer diameter of 160 mm × inner diameter of 153.5 mm × 50 mm tall, positioned at 250 mm below the hopper. A high-speed camera was used to record the flow of the material. The drained angle of repose was also measured after the discharge was completed.

Despite the capability of FBH to make an angle of repose Θ_R measurement, the pile formation was disrupted by the rapid and unsteady flow towards the end of the hopper discharge test. For better measurement consistency, the angle of repose was determined by the hollow cylinder method [10].

Development of a DEM Material Calibration Model

The interactions of "real-life" particles are complicated with large particle sizes, shapes, surface roughness and charge variances. Agglomeration of granular solids brings further complexities with particle arrangement in the granule, producing surface asperities, intragranular porosity, as well as different adhesion sources and other variables, such as uneven moisture distributions in materials [11]. If the contact model is to be applied at the single particle level, it is necessary to establish the model parameters at the actual particle-particle interaction level. This microstructure method would require considering a highly complex interface interaction situation, including surface topology, chemistry and interstitial media properties.

It is challenging to capture the complex microscale particle-particle contact. Therefore, bulk solid behavior is typically modelled on a mesoscale. With this methodology, each DEM particle represents the mesostructure of the bulk material. The particle-particle and particle-boundary interaction properties can be determined from bulk scale experiments, such as bulk density, compressibility and angle of repose, rather than measuring the microscopic interaction properties, like friction or adhesion surface energy for each particle.

The rationale for this is that DEM simulations are computationally demanding. To accurately model an existing system, it would be excessive to analyze every single particle at life-size, which makes this method prohibitive for any industrial-scale application. Previous studies have shown that even if the actual particle-particle interaction is measured accurately and applied to the contact model, it is still not adequate to simulate the bulk behavior of the material [12,13]. For the process development of industrial-scale applications, researchers are interested in the bulk material's flow behavior and handling characteristics. The purpose of the contact model is to emulate the interactions between the conglomerates to simulate the flow properties of the bulk material and the handling scenario of industrial applications.

In this paper, the Elasto-plastic Adhesion Model (EpAM) [14–18] was selected as the contact model. This contact model was coded to simulate the compressibility and adhesion of bulk materials. The scaling issue caused by the increase in particle size while keeping the size of the experimental apparatus the same is one of the primary sources of error during calibration. The selected particle size (Fig. 2b) is significant compared to the 10 mL Freeman FT-4 split vessel (Fig. 2a). Therefore, instead of replicating the experimental setup of Freeman FT-4 for calibration, compressibility (Fig. 5) and wall friction (Fig. 6) were determined by a Jenike shear tester [19], with the assumption that the material properties measured by these methods are similar and relevant.

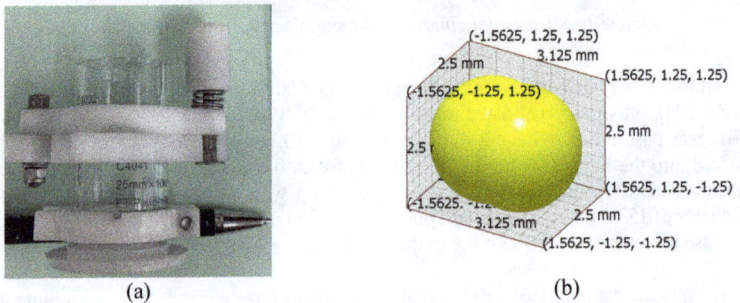

(a)   (b)

Figure 2: (a) Freeman FT-4 25mm × 10mL (Part# C4041) Split Vessel; (b) Bulk material is represented by multiple-spherical particles with a diameter of 2.5mm and an aspect ratio of 1.25.

Parameters were selected and adjusted according to a series of virtualized calibration tests processed by Altair EDEM® until realistic and accurate results were obtained. Calibration results are shown in Table 1, Fig. 3, Fig. 4, Fig. 5 and Fig. 6. For general reference, the specification of the custom-built desktop PC is summarized in Table 2. For the capability of running two simulations simultaneously or solving a large model with more than 1 million particles, this PC was configured to have two graphics processors.

Table 1: Calibrated bulk material properties

Parameter	Actual	DEM Simulation
Bulk Density	1.21 [g/cm^3]	1.21g/cm^3
Drained Angle of Repose (Fig. 3)	38-42°	40-41°
Static Angle of Repose (Fig. 4)	39-42°	40-42°

Table 2. Custom-built Desktop PC specification for DEM computational simulations

Operating System	Microsoft Windows 11® Pro 21H2
DEM Software	Altair EDEM v2022.1
Processor	12th Gen Intel® Core™ i7-12700F 2.10 GHz
Graphics Processor	Two × NVIDIA® GeForce® RTX 3080 Ti
Installed RAM	16.0 GB DDR4 3200 MHz
Hard Drive	2TB Crucial® CR2000MX

(a) (b)

Figure 3: Flat Bottom Hopper discharge test: (a) actual; (b) DEM simulation

(a) (b)

Figure 4: Static angle of repose: (a) actual; (b) DEM simulation

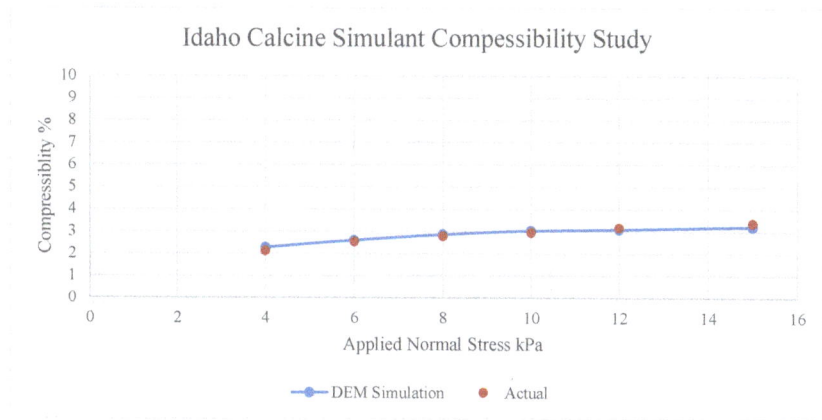

Figure 5: Compressibility of Idaho calcine simulant measured by Freeman FT4 (actual) and Jenike Shear Tester (DEM simulation).

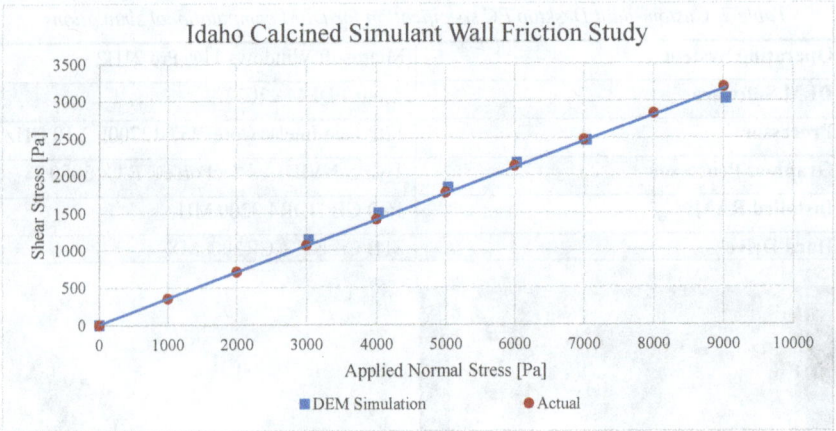

Figure 6: Idaho calcine simulant Wall Friction measured by Freeman FT4 (actual) and Jenike Shear Tester (DEM simulation).

Development of DEM HIP Canister Filling System

After obtaining the calibrated contact model, a three-dimensional DEM model was developed according to the geometry of an integrated HIP canister filling. Two bulk handling scenarios were simulated – standard operation and overfilled scenario.

In the standard operating scenario (Fig. 7), the powder is discharged from the hopper by opening an isolation butterfly valve into the inlet of a rotary valve. The rotary valve rotor (consisting of six pockets) was then rotated +/-180° for 40 cycles to discharge the waste simulant into a container via a unique patented filling nozzle-port coupling system [20]. Material quantity and the pile formation surface profile were examined.

The filling nozzle and process piping were initially filled with materials in the second simulation to simulate an overfill scenario (Fig. 9). After that, the filling nozzle-port coupling system was shut to stop powder flow. The container was then lowered to disconnect. This assessment investigated the amount of residual powder on the top of the filling port. A virtual false floor was placed inside the container, so less particle and simulation time was needed to achieve the overfilled scenario.

Results and Discussions

With the affordable hardware specified in Table 2, simulations of the bulk material handling scenarios were completed in less than five days. The importance of selecting an appropriate time step in DEM simulation was discussed in a previous study [21]. In this research, two identical simulations were set to run at 0.10 and 0.05 Euler time integration to better understand the effect of the time step on the simulated standard operating scenario. The amount of powder received in the container after 40 rotary valve powder dispensing cycles was 2.18276 kg with 0.10-time integration, whereas it was 2.18300 kg with 0.05-time integration. The discrepancy was negligible and within the measurement error of this experiment. As a result, the analysis of the current work continued to use 0.10 as the time step because it was adequate for this application.

In the experiment, 2.197 kg of powder was displaced by 40 rotary valve cycles, which showed that the model could accurately predict the amount of powder dispensed by the rotary valve with an error of 0.65%. Furthermore, the pile formation inside the container closely resembled the experimental observation. The predicted preferential filling towards the left side was verified, but with a steeper angle of repose than anticipated. Two additional 40-cycle fills were carried out with vibratory compaction between every 40 cycles to level the surface angle of repose of the pile. Results (Fig. 8) show that the developed DEM model can accurately predict mass flow and hence, cumulative or totalized mass.

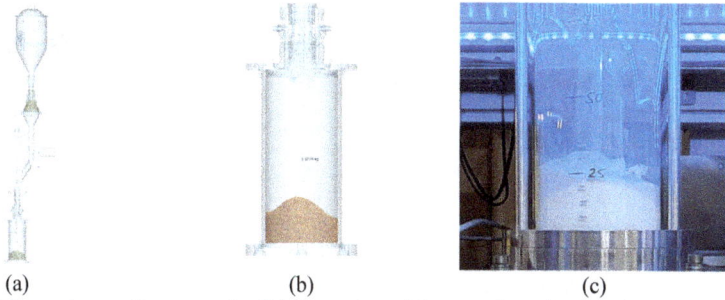

(a)   (b)   (c)

Figure 7: (a) Normal operating scenario; (b) Front view of the container showing the amount of powder received after 40 rotary valve powder dispensing cycles; (c) actual pile formation.

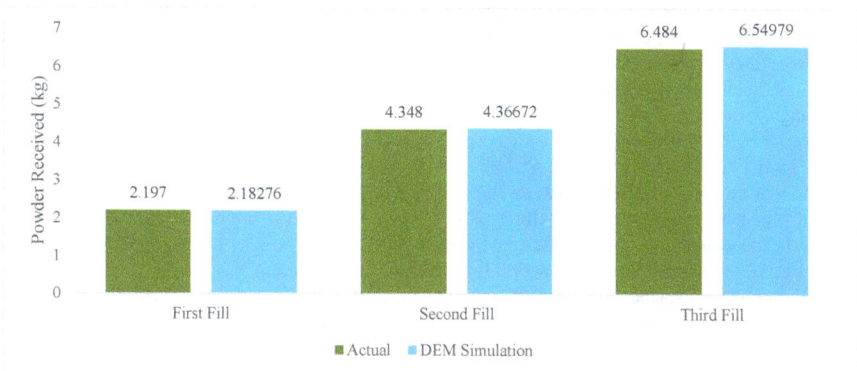

Figure 8: Mass received after 40 cycles (First Fill), 80 cycles (Second Fill) and 120 cycles (Third Fill) of rotary valve discharge cycles.

In the second simulation, the model successfully simulated the fail-safe feature of the filling system. The nozzle-port coupling system could stop the power flow without causing an overspill after removing the container. With a 0.10 time integration, 5.23056g of residual powder was accumulated on the top of the port. Compared to the 6g of residual powder found in the experiment, this is equivalent to an error of 12.8%.

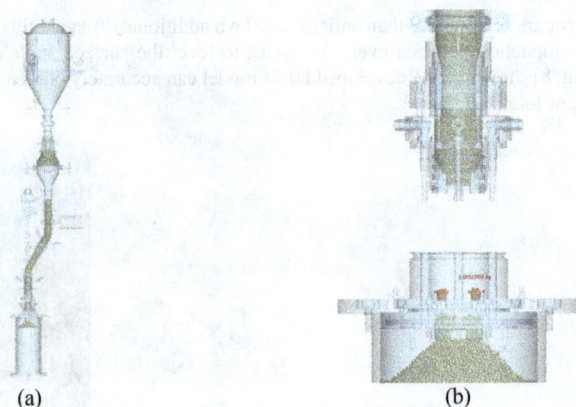

| (a) | (b) |

Figure 9: (a) Overfilling scenario; (b) Fail-safe powder containing disconnection

Conclusion

Predictions made by the calibrated DEM model on the amount of powder dispensed by the rotary valve are quite accurate. Optimization of the model can further reduce simulation time. Therefore, DEM simulation is also a powerful tool for optimizing existing processes. For example, a sensitivity study can be carried out to assess the impact of the pipeline routing on the surface profile of the pile formation. The issue of preferential filling can be resolved by engineering solutions, such as adjusting the filling nozzle's orientation, adding powder flow deflectors, or applying vibratory compaction to level the powder bed.

Powder hold-up above the HIP canister port after an overfilling scenario was predicted by the DEM model to be higher than expected. The overestimation is believed to be caused by the mechanical interlocking issue between the scaled particle and the nozzle outlet. Future research should repeat the experiment with finer particles to investigate if the discrepancy was an issue related to the scaled particle size. Despite the overestimation of powder hold-up, it is insignificant compared to the overall inventory within the powder filling system.

The simulated surface profile of the filled material in the container closely resembles experimental observation but with a steeper angle of repose than anticipated. Future research should consider an alternative calibration method to determine the angle of repose, such as discharging material from a funnel to form a poured angle of repose.

In-situ measurement is found to be challenging in the current experimental setup. Future development should consider non-disruptive measuring methods, such as LiDAR 3D scanning technology and a new container design, to better quantify the surface profile for model validations. The current research can also be extended to investigate the filling of HIP canisters with different geometries to support the development of near-net-shape manufacturing.

In conclusion, DEM simulation can accurately predict a HIP canister filling process when the contact model is adequately calibrated against powder characterization data. The predicted powder discharge quantity and surface profile of the pile formation are valuable design inputs for the preliminary engineering of HIP canister filling systems for producing advanced engineering materials and nuclear wasteforms. These design inputs are particularly beneficial for selecting process equipment and capacity assessment.

References

[1] M. W. A. Stewart, S. Moricca, P. G. Heath, S. Chung, and N. Hyatt, 'The Evolution of Hot-Isostatic Pressing for the Treatment of Radioactive Wastes - 18276', *WM Symposia, Inc.* United States, Jul. 01, 2018.

[2] M. D. Staiger and M. C. Swenson, 'Calcined Waste Storage at the Idaho Nuclear Technology and Engineering Center'. Idaho Cleanup Project Core, Idaho Falls, ID (United States), Sep. 24, 2018. https://doi.org/10.2172/1492726

[3] S. bin Yeom, E. Ha, M. Kim, S. H. Jeong, S. J. Hwang, and D. H. Choi, 'Application of the discrete element method for manufacturing process simulation in the pharmaceutical industry', *Pharmaceutics*, vol. 11, no. 8. MDPI AG, Aug. 01, 2019. https://doi.org/10.3390/pharmaceutics11080414

[4] M. Sakai, 'How should the discrete element method be applied in industrial systems?: A review', *KONA Powder and Particle Journal*, vol. 2016, no. 33. Hosokawa Powder Technology Foundation, pp. 169–178, 2016. https://doi.org/10.14356/kona.2016023

[5] W. R. Ketterhagen, M. T. Am Ende, and B. C. Hancock, 'Process modeling in the pharmaceutical industry using the discrete element method', *J Pharm Sci*, vol. 98, no. 2, pp. 442–470, 2009. https://doi.org/10.1002/jps.21466

[6] O. Baran, A. DeGennaro, E. Ramé, and A. Wilkinson, 'DEM simulation of a schulze ring shear tester', in *AIP Conference Proceedings*, 2009, vol. 1145, pp. 409–412. https://doi.org/10.1063/1.3179948

[7] B. A. Staples, G. S. Pomiak, and E. L. Wade, 'Properties of Radioactive Calcine Retrieved from the Second Calcined Solids Storage Facility at ICPP 4', Mar. 1979. Accessed: Oct. 12, 2022. [Online]. Available: https://www.osti.gov/servlets/purl/6071676

[8] 'ASTM D7891-15 Standard Test Method for Shear Testing of Powders Using the Freeman Technology FT4 Powder Rheometer Shear Cell', *ASTM International*, vol. 04.09. Dec. 27, 2016. https://doi.org/10.1520/D7891-15

[9] A. P. Grima, 'Quantifying and modelling mechanisms of flow in cohesionless and cohesive granular materials', Doctor of Philosophy thesis, University of Wollongong, 2011. [Online]. Available: http://ro.uow.edu.au/theses/3425

[10] H. M. Beakawi Al-Hashemi and O. S. Baghabra Al-Amoudi, 'A review on the angle of repose of granular materials', *Powder Technology*, vol. 330. Elsevier B.V., pp. 397–417, May 01, 2018. https://doi.org/10.1016/j.powtec.2018.02.003

[11] J. P. Morrissey, S. C. Thakur, and J. Y. Ooi, 'EDEM Contact Model: Adhesive Elasto-Plastic Model', Jun. 2014.

[12] T. A. H. Simons, R. Weiler, S. Strege, S. Bensmann, M. Schilling, and A. Kwade, 'A ring shear tester as calibration experiment for DEM simulations in agitated mixers - A sensitivity study', in *Procedia Engineering*, 2015, vol. 102, pp. 741–748. https://doi.org/10.1016/j.proeng.2015.01.178

[13] J. Quist and M. Evertsson, 'Framework for DEM Model Calibration and Validation', Sep. 2015.

[14] J. P. Morrissey, J. Y. Ooi, K. Tano, and G. Horrigmoe, 'An experimental and DEM study of the behaviour of iron ore fines', Sep. 2012.

[15] J. P. Morrissey, J. Y. Ooi, and J. F. Chen, 'A DEM study of silo discharge of a cohesive solid', in *In: Proceedings of Particle-Based Methods III Fundamentals and Applications - Particles*, 2013, pp. 298–309.

[16] J. P. Morrissey, J. Y. Ooi, J. F. Chen, K. Tano, and G. Horrigmoe, 'Experimental and discrete element modelling of cohesive iron ore fines', in *Proceedings of Particle-Based Methods III Fundamentals and Applications - Particles*, 2013, pp. 224–235.

[17] S. C. Thakur, J. P. Morrissey, J. Sun, J. F. Chen, and J. Y. Ooi, 'Micromechanical analysis of cohesive granular materials using the discrete element method with an adhesive elasto-plastic contact model', *Granul Matter*, vol. 16, no. 3, 2014. https://doi.org/10.1007/s10035-014-0506-4

[18] S. C. Thakur, H. Ahmadian, J. Sun, and J. Y. Ooi, 'An experimental and numerical study of packing, compression, and caking behaviour of detergent powders', *Particuology*, vol. 12, no. 1, 2014. https://doi.org/10.1016/j.partic.2013.06.009

[19] 'AS 3880:2017 Flow properties of coal'. Australian Standard, Aug. 30, 2017.

[20] S. Moricca and S. Chung, 'WO2018169594 RADIOACTIVE GRANULAR DISPENSING DEVICE'. https://patentscope.wipo.int/search/en/detail.jsf?docId=WO2018169594&_cid=P21-L7N63R-39264-1

[21] C. O'Sullivan and J. D. Bray, 'Selecting a suitable time step for discrete element simulations that use the central difference time integration scheme', in *Engineering Computations (Swansea, Wales)*, 2004, vol. 21, no. 2–4, pp. 278–303. https://doi.org/10.1108/02644400410519794

Materials Research Forum LLC
https://doi.org/10.21741/9781644902837-20

Computational Modeling of PM-HIP Capsule Filling and Consolidation by DEM-FEA Coupling

S. Sobhani[1,2,a*], M. Albert[3,b], D. Gandy[3,c], A. Tabei[1,2,d], A. Fan[1,2,e]

[1]School of Mechanical, Industrial, and Manufacturing Engineering, Oregon State University, Corvallis OR, USA

[2]Advanced Technology and Manufactruing Institute (ATAMI), Oregon State University, Corvallis OR, USA

[3]Electric Power Research Institute (EPRI), Charlotte NC, USA

[a]sobhanis@oregonstate.edu, [b]malbert@epri.com, [c]davgandy@epri.com, [d]ali.tabei@gmail.com, [e]zhaoyan.fan@oregonstate.edu

Keywords: Hot Isostatic Pressing (HIP), Modeling, Discrete Element Method, Finite Element Analysis, Capsule Filling

Abstract. Power Metallurgy Hot Isostatic Pressing (PM-HIP) is a manufacturing process, capable of producing net shape or near net shape components with complicated geometries from materials that are often difficult to cast and/or deform. However, the post-HIP quality and requirement of any additional process, such as machining, depends on the design and geometric complexity of the capsule. First of a kind geometry often requires several iterations of prototype builds. Considering the cost and long durations of HIP cycles, usage of computer models in order to predict parameters for an optimal capsule design of a PM-HIP process which produces a sound product in *the first trial* is extremely valuable. In this study, the pre-consolidation capsule filling process is simulated by Discrete Element Method (DEM) to capture the initial relative density. Finite element analysis (FEA) modeling of HIP, which includes a combined constitutive model based on compressive and consolidative mechanical behavior of powder uses the DEM results as input. Accuracy of the simulation tool is confirmed by comparing against a corresponding physical PM capsule fabrication and HIP experiment with pre- and post-HIP 3D scanning. The result shows that consolidation occurs as the model predicts, with negligible deviations on sharp edges.

Introduction

In HIP, a thin-walled capsule is filled with metal/alloy powder. After that the capsule is sealed and outgassed and then goes under a high-pressure and high-temperature HIP cycle for a given duration, usually in the order of 2 to 4 hours. The high pressures and temperatures lead to a noticeable shrinkage and deformation of the capsule [4,5] resulting in a near net-shaped part/component. A slight deviation in HIP parameters and factors can result in both microstructural and/or geometrical anomalies in the final produced part/component [6].

The filling step and early stages of the HIP process often involve the use of atomized metal powders; therefore, the Discrete Element Method (DEM) is the appropriate tool for modeling purposes [7]. At later stages of HIP where some degree of consolidation happens, FEA is more prominent in predicting the behavior and more applicable for to continuum models [7–13].

One of the more critical input parameters in the HIP process is the relative density (RD), the ratio of the density of the powder compact to the density of the bulk material. Accordingly, the central phenomenon of the HIP process is the evolution of RD to unity. Therefore, understanding RD distributions of the pre- and post-HIP parts are critical for a robust prediction of the shape of the final compact (solidified powder) and behavior of metal powders during the process [2, 3].

Hot Isostatic Pressing - HIP'22   Materials Research Forum LLC
Materials Research Proceedings 38 (2023) 141-149   https://doi.org/10.21741/9781644902837-20

According to the literature, the RD in filled HIP capsules may vary between 54% to 74% depending on the metal powder properties, filling procedure parameters, and the geometry of the capsule [14]. Previous studies have shown that establishment of a homogenous RD for the powder can help to achieve more accurate design of the component [8, 15]. The evolution of RD is a direct function of the volume change (consolidation/shrinkage), which is related to permanent deformation of the compact [16, 17]. Abouaf [5] developed an incremental and implicit finite element algorithm for compressing metal powders. It was proposed to extend the previous porous material plasticity theories of Oyane et. al. [18] and Kuhn [19] to capture complexities of HIP. The additional deformation in HIP was captured by accounting for creep, as the component is exposed to high temperatures.

Wickman et. al. [20] looked to develop a numerical simulation method to predict the final shape of the HIP'ed products by focusing on reducing the cost of the final parts by reducing the amount of machining required. Rate-dependent (viscoplastic) and rate-independent deformations were considered in this study. Based on dilatometry experiments, considerable deformation and particle rearrangement occurred during the early stages of HIP. However, due to fitting the viscoplastic deformation parameters to the entire process, the final axial deformation showed larger deformation. Van Nguyen and coworkers [8], discussed the importance of the density distribution, local RD gradient and knowledge of the amount of RD in improving the deformation and shrinkage during the HIP stages.

To investigate the effects of filling parameters on RD, three different capsule filling methods: 1) benign filling, 2) filling with tapping and 3) filling with vibration; were considered. The result suggested that near net shape parts require a stable powder particle size distribution, good mixing, and homogenous density gradient (close range of RD). Each of these are obtainable by conducting filling with tapping and vibrations.

In the current study, the pre-consolidation capsule filling process is simulated by discrete element method (DEM) to capture the initial relative density. Then FEA modeling of the HIP process with the DEM results as an input was used. This modeling includes a combined constitutive model based on mechanical behavior of the powder during compaction and consolidation in conjunction.

Methodology
1. Discrete Element Method (DEM)
DEM is a numerical technique based on a set of motion equations to model the movement and flow behavior of granular media [21,22]. DEM is founded on analyze forces due to particle-particle and particle-wall interaction [23,24]. The motion of each particle is captured by the Newton's second law. Therefore, for particle i in a body of powder, the following equations are derived:

$$m_i \frac{dv_i}{dt} = m_i g + \sum_{i=1}^{n} \left(F_{n,ij} + F_{t,ij} \right) \tag{1}$$

$$I_i \frac{d\omega_i}{dt} = \sum_{i=1}^{n} \left(T_{t,ij} + T_{r,ij} \right) \tag{2}$$

Where m_i is the mass of particle i and n represents the total number of particles in the system. v is velocity. F_t and F_n are respectively the tangential and normal contact force components. I_i is the moment of inertia of particle i, ω is angular velocity, and T_t, T_r are respectively the tangential and rolling torque [25]. In HIP process, particles have compressible plastic behavior, therefore Hysteretic Linear Spring Contact model is chosen to use in this work, due to its capability to show acceptable model for the granular material. Hysteretic Linear Spring Contact Model (HLSC) is based on the formalism proposed by Walton and Braun [26].

2. Constitutive model

For the FEA simulation used in this research, a combined constitutive model that can capture the complexities of the various active mechanisms in an acceptable way, is presented by Van Nguyen et. al. [29]. This combined model is based on the rate-dependent model by Abouaf [32] and Kuhn's and Downey's time-independent plasticity model [33]. In this model plastic yielding, linear isotropic hardening, and viscoplastic deformation lead to volume shrinkage and densification. The total strain developed in the powder is decomposed as follows:

$$\varepsilon_{ij} = \varepsilon_{ij}^{el} + \varepsilon_{ij}^{inel} + \varepsilon_{ij}^{th} \tag{3}$$

$$\varepsilon_{ij}^{inel} = \varepsilon_{ij}^{pl} + \varepsilon_{ij}^{cr} \tag{4}$$

Where ε_{ij} is total strain and $\varepsilon_{ij}^{el}, \varepsilon_{ij}^{inel}, \varepsilon_{ij}^{th}$ are elastic strain, inelastic strain and thermal strain. The inelastic strain is then decomposed into ε_{ij}^{pl} and ε_{ij}^{cr} which represent plastic and creep strains. Both primary and secondary creep mechanisms are considered in the creep strain term. Based on Kuhn's and Downey's work [19], an ellipsoidal yield surface which is a modified form of Von Mises criterion was used in the combined model. The yield function f is described by the following equations:

$$_j, \rho, \varepsilon_p) = \sigma_{eq} - r_1(\rho, p) - \sigma_y(\rho) = [A(\rho) J_2 + B(\rho) I_1^2]^{1/2} - h\rho^m - \sigma_0 \rho^k \tag{5}$$

Where σ_{eq} is the equivalent plastic stress that is a function of Cauchy stress components and the relative density via the $A(\rho)$ and $B(\rho)$, which are themselves functions of the relative density and the plastic Poisson ratio v_{pl}. In linear isotropic hardening, $r_1(\rho, p)$, which the relative density (ρ) and equivalent plastic strain (ε_p) [30] had an impact linear hardening for porous materials.

The tangential modulus h and exponent m are obtained by uniaxial compression experiments. $\sigma_y(\rho)$ is the initial yield stress which is a function of relative density, ρ. σ_0 is the yield stress in the fully dense material and k is a parameter obtained from detailed experiments.

Time independent inelastic deformation via creep, based on the Abouaf et. al. [32] creep model is used in this study. The creep rate ($\dot{\varepsilon}_{ij}^{cr}$) for porous materials is represented in the form of:

$$\dot{\varepsilon}_{ij}^{cr} = A(T) \sigma_{eq2}^{N(T)-1} \left(\frac{3}{2} c(\rho) S_{ij} + f(\rho) I_1 \delta_{ij} \right) \tag{6}$$

Where $A(T)$ is a function of temperature, $N(T)$ is the Dorn's constant parameter and $c(\rho)$ and $f(\rho)$ are experimental values measured as functions of relative density. σ_{eq} represented the equivalent stress due to creep deformation [34].

$$\sigma_{eq}^2 = 3c(\rho) J_2 + f(\rho) I_1^2 \tag{7}$$

Thermal strain increment is calculated as follows:

$$\Delta\varepsilon_{ij}^{th} = \alpha_{th}.\Delta T.\delta_{ij} \tag{8}$$

Where α_{th} is the thermal expansion coefficient that depends on relative density and temperature. ΔT and δ_{ij} are respectively, the change in temperature and the Kronecker's delta[29][15,34].

The density and temperature dependent model described above was coded in FORTRAN. As mentioned before, initial relative density values were transferred from DEM simulations; then input into the finite element analysis software ANSYS, through the User Programmable Feature (UPF) capability of ANSYS. ANSYS calculation engine solves for location dependent stresses and strains [35] for every timestep via Newton-Raphson iteration and updates density, ρ, accordingly based on the Eq. 8.

$$\rho^{t+\Delta t} = \rho^t \exp\left(\Delta\varepsilon_{kk}^{ine}\right) \tag{9}$$

3. Application of the Model

In the current study, the combined DEM and FEA models are used to simulate the HIP compaction of an example capsule geometry which can then be compared to experimental physical HIP capsule and 316L components. The application starts with investigating the initial density from the pre-consolidation process using DEM software. The 3D geometry used for the DEM part is shown in Fig. 1. (for the FEA simulation, a 2D axisymmetric figure is used to reduce the computational times since the component is symmetric around the Z axis). The particle distribution curve of the 316L powder is shown in Fig. 2. The pre-consolidation powder filling parameters such as time, and vibrational setups (frequency and amplitude) has been considered as 15 minutes, 60 HZ and 5 mm. The DEM filling simulation then provides an initial density that is transferred to the FEA model for stainless steel 316L. The material constants for the aforementioned equations are simulated with FORTAN code and the final dimensions of the capsule can be evaluated with experiment results.

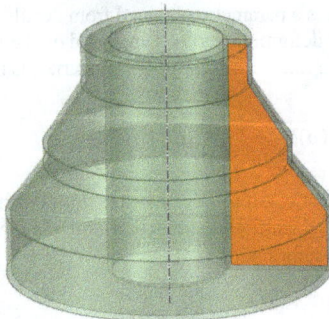

Figure 1. 3D model and 2D axisymmetric geometry

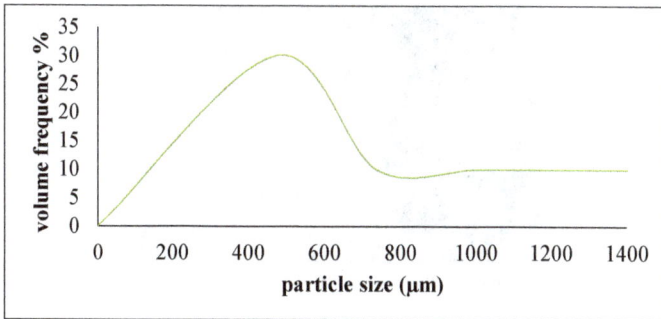

Figure 2. Particles size distribution

Results and Discussion

Figure 3 provides the outcome of DEM simulation of filling. Before the vibration starts (Fig. 3(a)) particles are segregated and separated. Larger particles mostly congregate at the inner sides due to gravity. As the vibration is begun, the particles with variety of sizes move and mix well together especially at the top region of the capsule. By longer duration of vibration, smaller particles migrate to the bottom of the capsule and improve the relative density (Fig. 3(b)) near the bottom of the capsule.

The initial RD (filled volume fraction) transferred from DEM to FEA is provided in Fig. 4 shown in discretized volumes. Figure 5 (a) shows the initial RD along with the density plot during the simulated HIP cycle compaction process. The result shows that maximum initial density is 0.6796 and the minimum density is 0.6016. Figure 5 (b) suggests that as the compact RD approaches 1, the part reaches the full densification level. As shown, the regions with more complex edges and corners demonstrate a lower RD. This suggests that the complexity of the capsule geometry and shape of the capsule affect the RD. The sharp corners and small dimension sections in both experiments and modeling may show slight deviation from full densification, due to constrained deformation/shrinkage in those areas. In parallel, a HIP capsule of the same pre-HIP capsule geometry was fabricated from mild carbon steel, filled with 316L powder, and captured pre- and post-HIP dimensions geometries via 3D laser scanning shows in Fig. 6. Comparison between FEA simulated post-HIP and experimental post-HIP dimensions are presented in Fig. 7, show acceptable result between simulation and experiment.

Figure 3. DEM pre-consolidation process

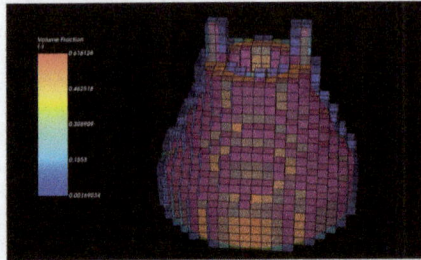

Figure 4. Volume fraction defined by Eulerian statistics exported from DEM as input to FEA

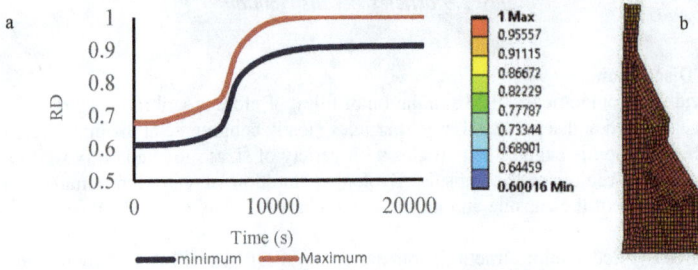

Figure 5. (a) RD plot during the HIP compaction cycle (b) final RD

Figure 6. Experimental capsule post-HIP (dots for 3D scanning)

Geometry	A	B	C	D	E	F
Post-HIP (FEM) (mm)	289.52	37.3	53.95	24.65	76.5	152.4
Post-HIP experiment (mm)	289.14	36.45	49.78	22.13	74.81	140.64
Deviation %	0.13	2.33	8.38	11.38	2.25	8.36

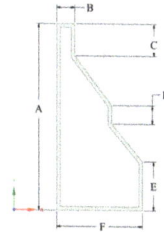

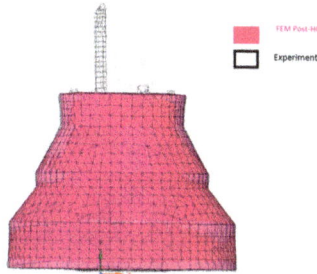

Figure 7. FEA modeling and HIP Experiment compared

Conclusions

In this study, the HIP process was modeled from the pre-consolidation stage to the final densified status. DEM was used to model the pre-consolidation filling process to establish the initial relative density (RD) of the compact, with the effects of vibration included. Finite element analysis (FEA) modeling of HIP, which includes a combined constitutive model based on compressive and consolidative mechanical behavior of powder uses the DEM results as input. The simulation predicted consolidation with an acceptable accuracy with the experiment. The result shows that full consolidation occurs with negligible deviations on sharp edges. Two additional example geometries are planned for fabrication and consolidation for further benchmarking of the simulation tool to real-world HIP component fabrication.

References

[1] Deng Y, Herzog S, Kaletsch A, Broeckmann C, Marmottant A, Laurent V. Numerical study of near-net-shape forming under encapsulation technologies and hip cladding. Proceedings Euro PM 2017: International Powder Metallurgy Congress and Exhibition 2017.

[2] Xue Y, Lang LH, Bu GL, Li L. Densification modeling of titanium alloy powder during hot isostatic pressing. Science of Sintering 2011;43:247–60. https://doi.org/10.2298/SOS1103247X

[3] Kim HS. Densification mechanisms during hot isostatic pressing of stainless steel powder compacts. Journal of Materials Processing Technology 2002;123:319–22. https://doi.org/10.1016/S0924-0136(02)00104-8

[4] Jeon YC, Kim KT. Near-net-shape forming of 316L stainless steel powder under hot isostatic pressing. International Journal of Mechanical Sciences 1999;41:815–30. https://doi.org/10.1016/S0020-7403(98)00053-8

Materials Research Forum LLC
https://doi.org/10.21741/9781644902837-20

[5]   Abouaf M, Chenot JL, Raisson G, Bauduin P. Finite element simulation of hot isostatic pressing of metal powders. International Journal for Numerical Methods in Engineering 1988;25:191–212. https://doi.org/10.1002/nme.1620250116

[6]   Redouani L, Boudrahem S. Hot isostatic pressing process simulation: Application to metal powders. Canadian Journal of Physics 2012;90:573–83. https://doi.org/10.1139/p2012-057

[7]   Abena A, Aristizabal M, Essa K. Comprehensive numerical modelling of the hot isostatic pressing of Ti-6Al-4V powder: From filling to consolidation. Advanced Powder Technology 2019;30:2451–63. https://doi.org/10.1016/j.apt.2019.07.011

[8]   Nguyen CV, Bezold A, Broeckmann C. Density distribution of powder in a HIP capsule after filling Density Distribution of Powder in a HIP Capsule after Filling 2011.

[9]   Martin CL, Bouvard D. Study of the cold compaction of composite powders by the discrete element method. Acta Materialia 2003;51:373–86. https://doi.org/10.1016/S1359-6454(02)00402-0

[10]   Deng Y, Kaletsch A, Bezold A, Broeckmann C. Precise Prediction of Near-Net-Shape HIP Components through DEM and FEM Modelling, 2019. https://doi.org/10.21741/9781644900031-24

[11]   Zhu HP, Zhou ZY, Yang RY, Yu AB. Discrete particle simulation of particulate systems : Theoretical developments 2007;62:3378–96. https://doi.org/10.1016/j.ces.2006.12.089

[12]   Elrakayby H, Kim HK, Hong SS, Kim KT. An investigation of densification behavior of nickel alloy powder during hot isostatic pressing. Advanced Powder Technology 2015;26:1314–8. https://doi.org/10.1016/j.apt.2015.07.005

[13]   Harthong B, Jérier J-F, Dorémus P, Imbault D, Donzé F-V. Modeling of high-density compaction of granular materials by the Discrete Element Method. International Journal of Solids and Structures 2009;46:3357–64. https://doi.org/10.1016/j.ijsolstr.2009.05.008

[14]   Van Nguyen C, Bezold A, Broeckmann C. Influence of initial powder distribution after pre-densification on the consolidation of stainless steel 316L during HIP. International Powder Metallurgy Congress and Exhibition, Euro PM 2013 2013.

[15]   Van Nguyen C, Bezold A, Broeckmann C. Anisotropic shrinkage during hip of encapsulated powder. Journal of Materials Processing Technology 2015;226:134–45. https://doi.org/10.1016/j.jmatprotec.2015.06.037

[16]   Khoei AR, Molaeinia Z, Keshavarz Sh. Modeling of hot isostatic pressing of metal powder with temperature–dependent cap plasticity model. Int J Mater Form 2013;6:363–76. https://doi.org/10.1007/s12289-012-1091-x

[17]   Chung SH, Park H, Jeon KD, Kim KT, Hwang SM. An optimal container design for metal powder under hot isostatic pressing. Journal of Engineering Materials and Technology, Transactions of the ASME 2001;123:234–9. https://doi.org/10.1115/1.1354992

[18]   Oyane M, Shima S, Kono Y. Theory of Plasticity for Porous Metals. Bulletin of JSME 1973;16:1254–62. https://doi.org/10.1299/jsme1958.16.1254

[19]   Kuhn HA, Downey CL. Material Behavior in Powder Preform Forging. ASME Pap 1972.

[20]   Wikman B. Modelling and Simulation of Powder Pressing with Consideration of Friction Modelling and Simulation of Powder Pressing with Consideration of Friction 1999.

Materials Research Forum LLC
https://doi.org/10.21741/9781644902837-20

[21]   Nakano M, Abe T, Kano J, Kunitomo K. DEM Analysis on Size Segregation in Feed Bed of Sintering Machine 2012;52:1559–64.

[22]   Alian M, Ein-mozaffari F, Upreti SR. Analysis of the mixing of solid particles in a plowshare mixer via discrete element method (DEM). Powder Technology 2015;274:77–87. https://doi.org/10.1016/j.powtec.2015.01.012

[23]   Cundall_Strack.pdf. n.d.

[24]   Zhou YC, Wright BD, Yang RY, Xu BH, Yu AB. Rolling friction in the dynamic simulation of sandpile formation 1999;269:536–53.

[25]   Qiao J, Duan C, Dong K, Wang W, Jiang H, Zhu H, et al. DEM study of segregation degree and velocity of binary granular mixtures subject to vibration. Powder Technology 2021;382:107–17. https://doi.org/10.1016/j.powtec.2020.12.064

[26]   Walton OR, Braun RL, Walton OR, Braun RL. inelastic , frictional disks Viscosity , Granular-Temperature , and Stress Calculations for Shearing Assemblies of Inelastic , Frictional Disks * 2013;949. https://doi.org/10.1122/1.549893

[27]   Chaudhuri B, Mehrotra A, Muzzio FJ, Tomassone MS. Cohesive effects in powder mixing in a tumbling blender 2006;165:105–14. https://doi.org/10.1016/j.powtec.2006.04.001

[28]   Stevens AB, Hrenya CM. Comparison of soft-sphere models to measurements of collision properties during normal impacts 2005;154:99–109. https://doi.org/10.1016/j.powtec.2005.04.033

[29]   Van Nguyen C, Deng Y, Bezold A, Broeckmann C. A combined model to simulate the powder densification and shape changes during hot isostatic pressing. Computer Methods in Applied Mechanics and Engineering 2017;315:302–15. https://doi.org/10.1016/j.cma.2016.10.033

[30]   Deng Y, Birke C, Rajaei A, Kaletsch A, Broeckmann C. Numerical Study of Hot Isostatic Pressing with Integrated Heat Treatment of PM-HIP Cold Work Steel D7. WorldPM 2018 2018:442–53.

[31]   Aryanpour G, Mashl S, Warke V. Elastoplastic-viscoplastic modelling of metal powder compaction: Application to hot isostatic pressing. Powder Metallurgy 2013;56:14–23. https://doi.org/10.1179/1743290112Y.0000000027

[32]   Abouaf M, Chenot JL, Raisson G, Bauduin P. Finite element simulation of hot isostatic pressing of metal powders. International Journal for Numerical Methods in Engineering 1988;25:191–212. https://doi.org/10.1002/nme.1620250116

[33]   Kuhn HA, Downey CL. Material Behavior in Powder Preform Forging. ASME Pap 1972.

[34]   Kohar CP, Martin É, Connolly DS, Patil S, Krutz N, Wei D, et al. A new and efficient thermo-elasto-viscoplastic numerical implementation for implicit finite element simulations of powder metals: An application to hot isostatic pressing. International Journal of Mechanical Sciences 2019;155:222–34. https://doi.org/10.1016/j.ijmecsci.2019.01.046

[35]   Group M. ANSYS USER Material Subroutine USERMAT. Technology 1999:1–22.

Identification of Porous Materials Rheological Coefficient Using Experimental Determination of the Radial and Longitudinal Strain Rate Ratio

Gerard Raisson[1,a*], Vassily Goloveshkin[2,b], Victor Samarov[3,c]

[1]Consultant, Cournon D'Auvergne, France

[2]IAM RAS, RTU MIREA, Moscow, Russia

[3]LNT PM Inc., Garden Grove, CA, USA

[a]gerard.raisson@gmail.com, [b]nikshevolog@yandex.ru, [c]victor@synertechpm.com

Keywords: HIP Modeling, Powder Rheology

Abstract. Traditional approach of identification of the f1 and f2 function of the Green plasticity criterion [1] for porous materials is based on the two experiments: isostatic pressing of samples in the interrupted cycles (determining density as a function of pressure providing the f2 values) and upsetting providing an equation for determining the f1 values responsible for the shear deformations. Treatment of HIP dilatometer data allows to get an independent f1 evaluation which shows that upsetting is not representative of stress state met during powder HIP densification. In the second part, a new approach is suggested allowing to determine these two functions using only the deviations from the isotropic shrinkage obtained in the interrupted cycles. This anisotropy is characterized by the ψ coefficient ratio between the radial and the axial strain rate.

Introduction

Traditional approach of identification of the f1 and f2 function of the Green plasticity criterion [1] for porous materials is based on the two experiments: isostatic pressing of samples in the interrupted cycles determining density as a function of pressure providing the f2 values and upsetting providing equations for determining the f1 values responsible for the shear deformations.

In [2], HIP dilatometer was presented (device and treatment of results). It was established an equation giving the energy equilibrium between effect of HIP pressure acting on external surface of capsule versus energy needed for strain of capsule and powder (densification and shear stress strain).

Modeling

Mechanical analysis

The basic Green equation is:

$$Y^2 = \frac{S_1^2}{9*f_2^2} + \frac{3}{2} * \frac{S_2^2}{f_1^2} \tag{1}$$

where Y is the flow stress of fully dense powder material and S1 and S2 are the two invariants of

stress tensor with $S_1 = \frac{\sigma_r + \sigma_\varphi + \sigma_z}{3}$ and $S_2 = \sqrt{\frac{1}{3} * \left((\sigma_r - \sigma_\varphi)^2 + (\sigma_z - \sigma_\varphi)^2 + (\sigma_z - \sigma_r)^2 \right)}$.

a. *HIP Dilatometer Experiments*

In HIP dilatometer experiments the equation for the energy equilibrium gives the following equation for the HIP pressure:

$$P = \frac{1}{(\varepsilon + 2A)} \left\{ \frac{Y}{3} \sqrt{\left(9f_2^2 - 2f_1^2\right)(2A + \varepsilon)^2 + 6f_1^2\left(2A^2 + \varepsilon^2\right)} + \right.$$
$$\left. + 2\frac{h}{R}T\sqrt{\frac{1}{3}}\sqrt{3\varepsilon^2 + (2A + \varepsilon)^2} \right\} \tag{2}$$

Y is the flow stress of full dense powder material, T is the flow stress of capsule material, P the HIP pressure on the capsule, h is the capsule thickness, R is the radius of capsule, f1 and f2 are coefficients defined in Green equation, ε is the radial strain rate and A the axial strain rate.

It is interesting to introduce coefficient ψ such as $\psi = \dfrac{\left(\dfrac{\Delta b}{b}\right)}{\left(\dfrac{\Delta a}{a}\right)}$ where Δb is the relative momentary axial shrinkage, b- the current height of powder layer, Δa - is the relative (momentary) radial shrinkage of powder and a - the current radius of powder. Equation 2 can be re-written as:

$$P = \left\{ \frac{Y}{3} \sqrt{\left(9f_2^2 - 2f_1^2\right) + 6f_1^2\frac{\left(2 + \psi^2\right)}{(\psi + 2)^2}} + \right.$$
$$\left. + 2\frac{h}{R}T\sqrt{\frac{1}{3}}\sqrt{3\frac{\psi^2}{(\psi + 2)^2} + 1} \right\} \tag{3}$$

$$\frac{dP}{d\psi} = Y\frac{-4f_1^2\dfrac{(1 - \psi)}{(\psi + 2)^3}}{\sqrt{\left(9f_2^2 - 2f_1^2\right) + 6f_1^2\dfrac{\left(2 + \psi^2\right)}{(\psi + 2)^2}}} + \frac{h}{R}T\sqrt{\frac{1}{3}}\frac{12\dfrac{\psi}{(\psi + 2)^3}}{\sqrt{3\dfrac{\psi^2}{(\psi + 2)^2} + 1}} \cdot \tag{4}$$

From the condition $\dfrac{dP}{d\psi} = 0$, we get:

$$\frac{h}{R}T\sqrt{3}\frac{\psi}{\sqrt{3\dfrac{\psi^2}{(\psi + 2)^2} + 1}} = Y\frac{f_1^2(1 - \psi)}{\sqrt{\left(9f_2^2 - 2f_1^2\right) + 6f_1^2\dfrac{\left(2 + \psi^2\right)}{(\psi + 2)^2}}} \cdot \tag{5}$$

Combining with equation (3), we get:

$$\frac{Y}{3}\sqrt{\left(9f_2^2 - 2f_1^2\right) + 6f_1^2\frac{\left(2 + \psi^2\right)}{(\psi + 2)^2}} + 2\frac{h}{R}T\sqrt{\frac{1}{3}}\sqrt{3\frac{\psi^2}{(\psi + 2)^2} + 1} = P \tag{6}$$

Then
$$\sqrt{\left(9f_2^2 - 2f_1^2\right) + 6f_1^2 \frac{\left(2+\psi^2\right)}{\left(\psi+2\right)^2}} = \frac{3\left[P - 2\frac{h}{R}T\sqrt{\frac{1}{3}}\sqrt{3\frac{\psi^2}{\left(\psi+2\right)^2}+1}\right]}{Y} \tag{7}$$

Combining with equation (5) we get:

$$\frac{h}{R}T\sqrt{3}\frac{\psi}{\sqrt{3\frac{\psi^2}{\left(\psi+2\right)^2}+1}} = Y^2 \frac{f_1^2\left(1-\psi\right)}{3\left[P - 2\frac{h}{R}T\sqrt{\frac{1}{3}}\sqrt{3\frac{\psi^2}{\left(\psi+2\right)^2}+1}\right]} \tag{8}$$

Then:

$$f_1^2 = \frac{h}{R}\frac{T}{Y^2}\left[P - 2\frac{h}{R}T\sqrt{\frac{1}{3}}\sqrt{3\frac{\psi^2}{\left(\psi+2\right)^2}+1}\right]3\sqrt{3}\frac{\psi}{\left(1-\psi\right)\sqrt{3\frac{\psi^2}{\left(\psi+2\right)^2}+1}} \tag{9}$$

We can get f2 using equations (6) and (9):

$$f_2^2 = \frac{1}{Y^2}\left[P - 2\frac{h}{R}T\sqrt{\frac{1}{3}}\sqrt{3\frac{\psi^2}{\left(\psi+2\right)^2}+1}\right]^2 - \frac{4}{9}f_1^2\frac{\left(1-\psi\right)^2}{\left(\psi+2\right)^2}. \tag{10}$$

b. *Evaluation of ψ coefficient*

Coefficient ψ was evaluated using HIP dilatometer results for TA6-4 powder HIPed in 304L capsule [2]. Theoretically ψ is only a function of powder relative density with following particularities:

    1. For initial density, f1=0 then ψ=0
    2. For full density, f1=1 then ψ has a given value depending on used rheology (0 for a purely viscous law and around 0.2 for an elasto-viscous law).

Figure 1 gives the values of ψ function of ρ for the bar trial based on the dilatometer measurements for:
    1. a very small step (20s)
    2. a 2% density step

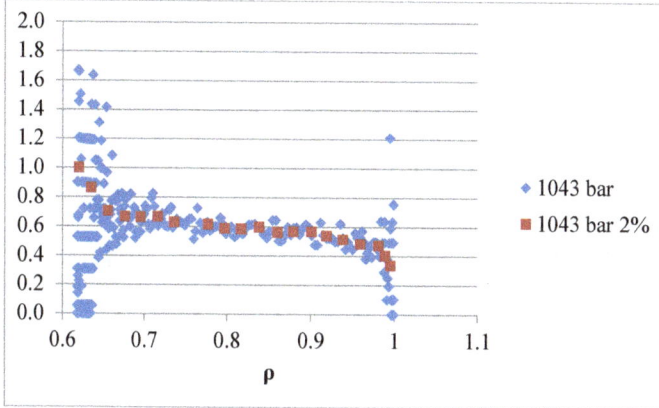

Figure 1: ψ function of ρ for the 1043 bar trial.

For medium density, evolution of ψ is rather even.

At low density, tendency is not compatible with what as expected. Even if a strong scatter is generated by low values of densification, it is necessary to involve the fact that used capsule was not really what generates instabilities of radial shrinkage due to the deformation of lids. Data are not valuable at low density. After some densification (around 0.7-0.75) flow pattern becomes stable, and data become valuable.

At high density, even if a strong scatter is generated by low densification rate, tendency seems compatible with expected evolution.

c. Identification of f1 and f2

Identification requires use of equations 8 and 9. However it is necessary to fix used rheologies for full dense powder material and capsule material. In fact, f1 and f2 cannot be identified by themselves, only a set of values (f1, f2, Y and T) can be identified. For example, in Fig. 2, are given flow stress of full dense powder material function of density with a viscous behavior [3] and for an elastic-viscous behavior [4]. Discrepancy is small but not negligible.

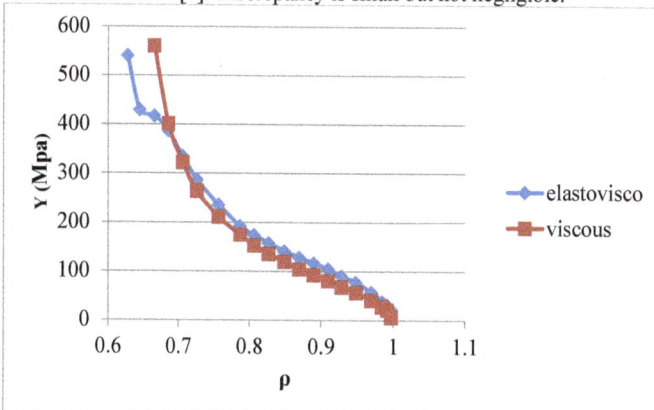

Figure 2: Flow stress of TA6-4 function of density along densification.

Materials Research Forum LLC
https://doi.org/10.21741/9781644902837-21

For capsule material rheology, an elastic-viscous- plastic law is used [5].
Figure 3 shows estimation of coefficient f1 with an elastic-viscous rheology for powder material (TA6-4).

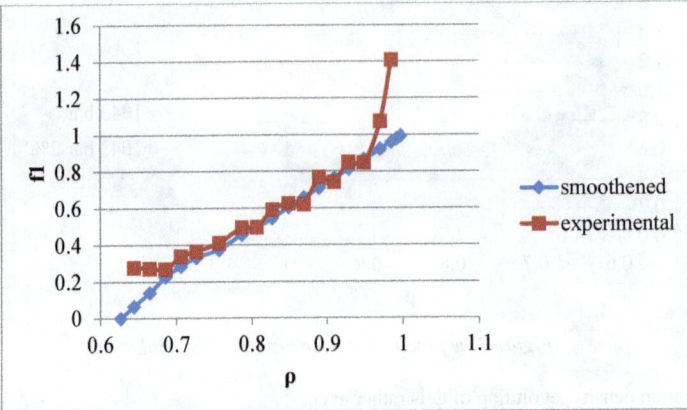

Figure 3: f1 function of density (experimental and after smoothening)

For medium density (between 0.7 and 0.95) variation is even and almost linear. At low density the measured value of f1 does not tend toward 0 and at high density does not tend toward 1.

At low density, it has been said that experimental data are not valuable. At high density, uncertainty due to scatter and very high sensitivity of f1 to ψ value explain this anomaly. An extrapolation of central part of the curve toward f1=0 for filling density and f1=1 for full density has been proposed. From proposed f1=f(ρ) curve, it is easy to calculate a new ψ=g(ρ) curve (see figure 4). The agreement with experimental points is satisfactory. It has been added ψ function of density with a purely viscous law for TA6-4.

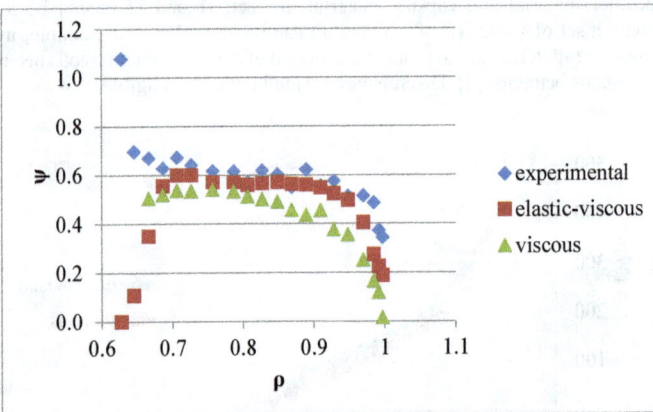

Figure 4: ψ function of density calculated using smoothened f1 =f(ρ) curve for two rheologies.

Figure 5 gives f2 values as a function of density. It can be noticed that these values are almost of those obtained in isotropic conditions. It means that $f_2 = \frac{P_{int}}{Y}$ where P_{int} is the pressure in the

Hot Isostatic Pressing - HIP'22   Materials Research Forum LLC
Materials Research Proceedings 38 (2023) 150-159   https://doi.org/10.21741/9781644902837-21

capsule with correction of pressure drop due to the wall of capsule. This result is very good, because it means that identification of f2 with the method of interrupted HIP cycle in a cylindrical capsule is valuable.

Figure 6 gives f2 values as a function of density for an elastic-viscous rheology and for a viscous law. The discrepancy expresses flow stress discrepancy.

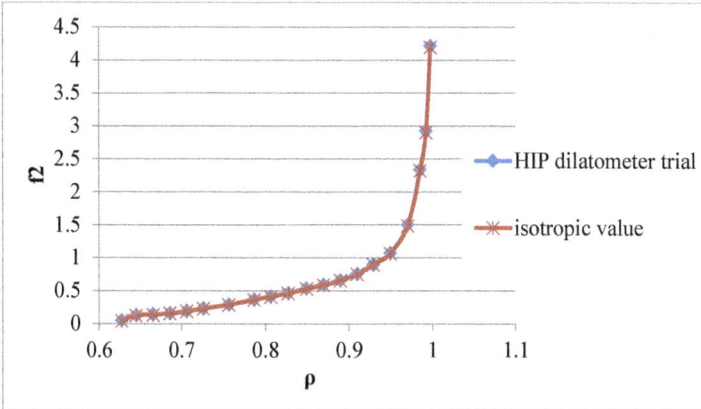

Figure 5: f2 function of density (HIP dilatometer data treatment and isotropic calculation).

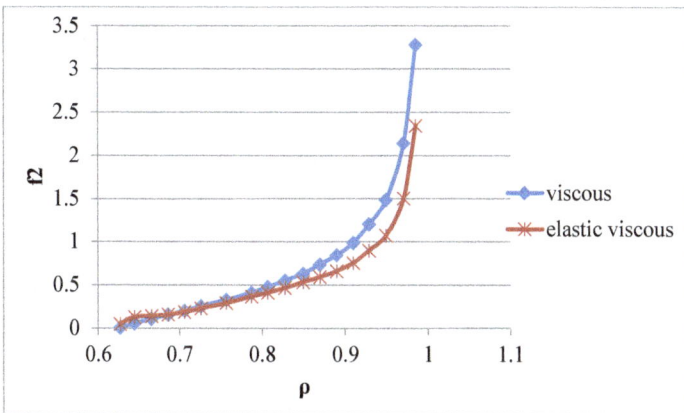

Figure 6: f2 function of density for viscous and elastic viscous law for TA6-4

d. Comparison with other identification processes

It is interesting to compare results of identification by exploitation of HIP dilatometer data with other existing techniques. Two are available:

1. Upsetting of partially dense blanks: for this process $f_1^2 = \dfrac{\sigma_z^2}{Y^2 - \dfrac{\sigma_z^2}{9*f_2^2}}$ where σ_z is the pressure

applied to the sample with a controlled strain rate. Two set of data have been used [3] and [6]

2. Unidirectional pressing of powder in a rigid die at a controlled strain rate with a measure of axial and axial stress: for this process [5], it was observed a ratio K between radial and axial stresses depending only on density.

$$f_1^2 = \frac{2*(1-K)^2*\sigma_z^2}{\sigma_{eq}^2 \frac{1}{9*f_2^2}*(1-2*K)^2*\sigma_z^2} \text{ with } K = \frac{\sigma_r}{\sigma_z} = 1 - 1.6*(1-\rho)^{0.45} \text{ for } \rho>0.75.$$

As said previously it is important to remember that f1 is not an independent parameter. Actually, it is an established set of data (full dense powder rheology, capsule material rheology, f1 and f2). It is necessary to homogenize data and sometimes it is not trivial.

For upsetting we have used [3] at 800°C and [4] at 925. For unidirectional pressing, we have used [3] at 800°C.

For TA6-4, we have used elastic-viscous rheology of [4]

For 304L (capsule material) we have used elastic-viscous-plastic rheology of CEA [5]

For f2, we have used Fig. 5 relationship.

Figure 7 gives f1 curves as a function of density for treated data.

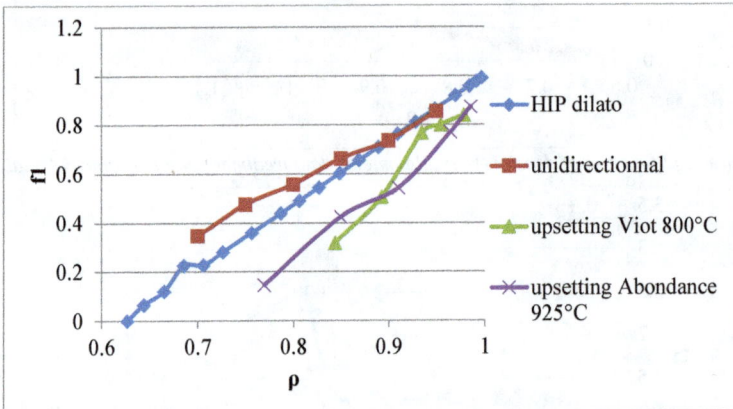

Figure 7: f1 function of density according to different identification processes.

If we place stress conditions of experimental points on the plasticity ellipse with $\frac{S_1}{\sqrt{3}}/Y$ and S_2/Y reduced coordinates for density levels calculated with HIP dilatometer data (TA6-4, f1 and f2) it can be noticed (Fig. 8 and 9) that:

Unidirectional pressing points are close of plasticity ellipse knowing that precision of data is limited particularly for HIP dilatometer.

In opposite upsetting points are definitively apart.

It is interesting to look at angular position of plasticity points: $\theta = Arctg(\frac{S_2}{S_1*\sqrt{3}})$ (Fig. 9)

For upsetting, positioning angle is constant: $tg(\theta) = \sqrt{2}$

For unidirectional pressing positioning angle varies in a large range.

For HIP dilatometer, angles are small which means that stress conditions are close of isotropic. An evaluation of angle range function of ψ range from 0 (flat densification) to 1 (isotropic densification) is between 18 and 0°. This angle has been put in Fig. 9.

Hot Isostatic Pressing - HIP'22   Materials Research Forum LLC
Materials Research Proceedings 38 (2023) 150-159   https://doi.org/10.21741/9781644902837-21

In HIP densification in a capsule, stress conditions are located between isotropic and flat densification HIP dilatometer according to the capsule design. Upsetting conditions are usually not met in regular HIP of capsules with powder.

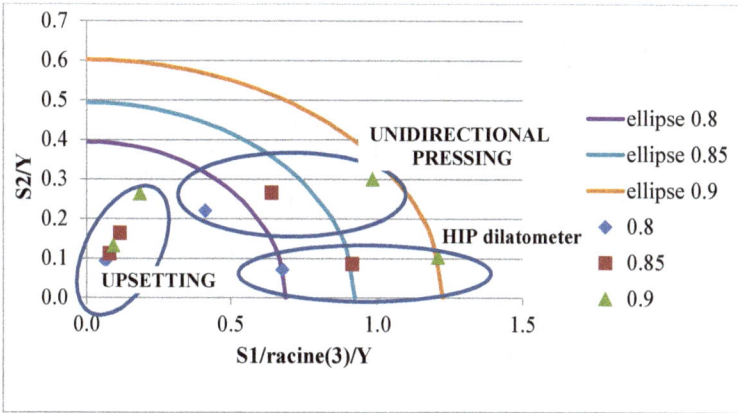

Figure 8: Reduced plasticity ellipse for 3 density levels and corresponding experimental points obtained with 3 identification processes.

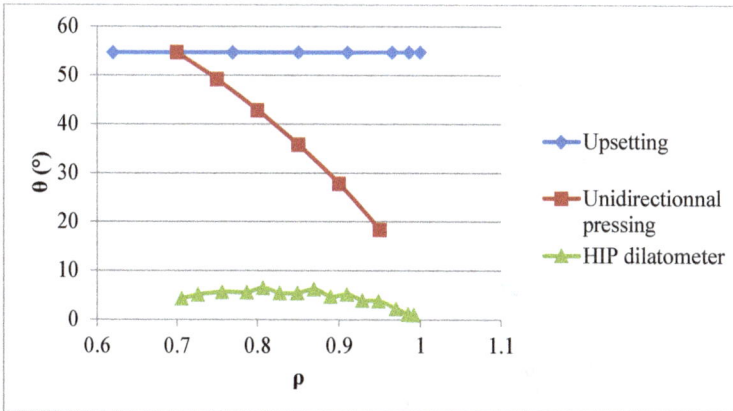

Figure 9: Angular positioning of experimental points in function of density and identification process.

Intermediate conclusions

Identification process of coefficients of Green's law is an important issue. Knowing that real stress conditions during HIP densification are located between 0 (isotropic) and 18° angles on the plasticity ellipse   and so, it is possible to claim that upsetting is not representative of met conditions. HIP dilatometer and unidirectional pressing gives closer results. It means that other efficient identification processes are existing and that it would not be useful to modify Green's law in order to introduce a coupling of isotropic and shear stresses as several authors have proposed [6, 7, 8].

Optimizations of determination of powder radial and longitudinal strain rates

Experimental conditions of HIP dilatometer are not optimal particularly at low density. There are two possibilities to improve the situation:

To design a new experimental HIP dilatometer device to use long thin wall capsules. It means investment and time.

To use classical interrupted cycles with optimized capsule geometry to generate a non-negligible shear stress and to get uniform axial strain in the middle of capsule where deformation pattern is uniform on a section.

The evenness of f1 function of density shows that identification between 0.75 and 0.9 density and a step of 5% (in order to limit temperature gap) which means 4 or 5 capsules would be efficient. Using available experimental data used previously, it has been carried out a simulation of a 5% density step in comparison of 2% step used has been done (figure 10). It shows that results are very similar and extrapolation toward tap density and full density is easy.

Measuring radial strain rate is easy after cutting and measuring of powder radius. Measuring axial strain rate either by a precise measurement of powder radius and capsule thickness after HIP using mass conservation equations or by a direct measurement of grooves drawn at the external surface of capsules.

A critical issue is to use the same temperature and pressure ramps in order to avoid scattering. It will be necessary to optimize capsule design in order to get a sufficient shear stress level in the capsule.

For identification of f2, the new proposed method must be more thoroughly analyzed for low density in order to account for thermal conductivity, which is a fundamental issue in the beginning of HIP cycle for thick parts.

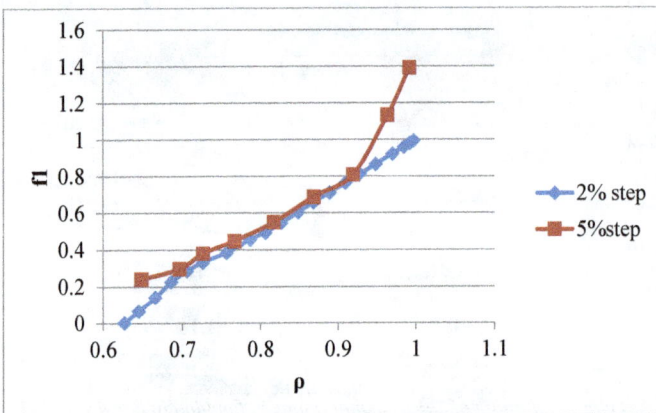

Figure 10: f1 function of density with 2 and 5% density step.

Conclusion

Comparison of experimental data of 3 identification processes shows that unidirectional pressing and HIP dilatometer gives consistent results for f1 coefficient. In opposite, upsetting conditions are not representative of met conditions during real powder densification.

It is possible to identify f1 coefficient with a better precision through a simplified process consisting in 5 or 6 HIP interrupted HIP cycles of cylindrical capsule with a precise dimensional

Materials Research Forum LLC
https://doi.org/10.21741/9781644902837-21

measurement before HIP and after HIP to reveal the difference between the axial and radial strain rates.

References

[1] R.J. Green: Int.Mech.Sci., 1972, 14, (4), 215-224. https://doi.org/10.1016/0020-7403(72)90063-X

[2] G. Raisson, Y. Kozyrev, A. Ponomarev, A. Bochkov, V. Goloveshkin: HIP 2017 proceedings

[3] D. Abondance: Thesis Universite Joseph Fourier Grenoble 1996

[4] Burlet, Gillia: Model identification for powder densification. HIP 2005 Paris May 2005.

[5] CEA Private communication

[6] P. Viot: PHD Thesis Universite Joseph Fourier Grenoble 2000

[7] C.L. Martin, D. Favier, M. Suery: International Journal of Plasticity, vol.15 pp.981-1008, 1999. https://doi.org/10.1016/S0749-6419(99)00017-0

[8] P. Stutz : Radenkovic Conference Palaiseau (Paris) 1992

Materials Research Forum LLC
https://doi.org/10.21741/9781644902837-22

A New Constitutive Modeling for Hot Isostatic Pressing of Powders

Gholamreza Aryanpour

University of Quebec at Chicoutimi, Chicoutimi, Quebec, Canada

Gholamreza_Aryanpour@uqac.ca

Keywords: Modeling, Plasticity, Viscoplasticity, Densification, Stainless Steel 316L

Abstract. Regarding the fact that a major deformation and thus densification of powder takes place during both ramp and holding stages in the process of hot isostatic pressing, a more precise model must be applied to describe the inelastic behavior of powder. For this aim, a plastic-viscoplastic modeling is developed in this work. This modeling is based on considering two logical phenomena, one is the strain hardening of powder particles during isostatic pressing and the other one is the increase in relative density following the particle deformation. Based on these facts, the plastic-viscoplastic model is formulated for the isostatic pressing. Finally, the numerical and experimental results will be compared and discussed.

Introduction

Hot Isostatic Pressing (HIP) is a sustainable manufacturing process to produce near net/net shape products in small to large sizes. The powder poured in a thin metallic container is compacted by isostatic pressure of an inert gas through applying some cycles of temperature and pressure. Fig. 1 is a schematic demonstration of the process.

In a general consideration, material inelastic deformation and thus densification take place at both ramp and holding stages shown in Fig. 1. From a phenomenological point of view, a pure plastic model is not able to predict any densification at the holding stage because pressure is held constant at this stage. On the other hand, an ideal viscoplastic model is not capable to describe material deformation at the ramp stage if the pressure ramp occurs at low temperatures. Therefore, a more complete modeling other than pure plastic or ideal viscoplastic model is needed. To achieve such a model, there are two solutions, either a unified viscoplastic or a plastic-viscoplastic model. In this research, we select the plastic-viscoplastic approach due to this fact that the parameters introduced in this modeling are relatively easy to be identified by some classical experiments.

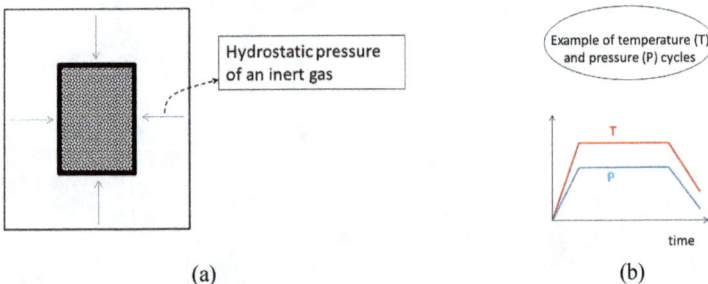

Figure 1. Schematic illustration of HIP process and example of temperature and pressure cycles.

Theoretical Basis of the Modeling

In a powder compaction process, the inelastic deformation of particles is responsible for material densification. Evidently, density or relative density (density of porous material to the density of the same material without porosity) is an isotropic hardening parameter. However, along with

Materials Research Forum LLC
https://doi.org/10.21741/9781644902837-22

densification, particle strain hardening should be considered as another isotropic hardening parameter if compaction is assumed in a general condition of temperature and stress. Before presenting the final equation of the model, the material strain hardening is clarified in the following.

Effect of Strain Hardening. Classical theory for material deformation in a uniaxial test considers an initial elastic domain before achieving an elastic limit beyond which a permanent deformation begins to happen. However, according to the microplastic phenomena that may occur during deformation, one can consider a plastic deformation from the beginning of loading so that the yield stress (σ_y) can be expressed as the following:

$$\sigma_y = k(\varepsilon^p)^m. \tag{1}$$

Where ε^p is the axial plastic strain and k and m are two material parameters. For a multiaxial loading, the material yield point is defined by $k\left(\varepsilon_{eq}^p\right)^m$ where ε_{eq}^p is the equivalent plastic strain defined by the following:

$$\varepsilon_{eq}^p = \int \left[\frac{2}{3}\dot{\varepsilon}_{ij}\dot{\varepsilon}_{ij}\right]^{\frac{1}{2}} dt. \tag{2}$$

Plasticity of Powder and Yield Function. As a powder is porous, the yield function is dependent on the first invariant of stress in addition to the second invariant of deviatoric stress. Therefore, regarding Eq. 1, the yield function given by Green [1] can be extended to the following form:

$$\left(hS_1^2 + \frac{3}{2}g\bar{S}_2^2\right)^{\frac{1}{2}} - k\varepsilon_{eq}^p{}^m = 0. \tag{3}$$

Where S_1 is the first invariant of stress tensor and $\bar{S}_2$ is equal to $\left[\bar{\sigma}_{ij}\bar{\sigma}_{ij}\right]^{\frac{1}{2}}$. $\bar{\sigma}_{ij}$ is the component of deviatoric stress. Function h can be defined as a function of relative density (ρ) with α and β as two constants.

$$h = \alpha \left(\frac{1-\rho}{\rho-\rho_0}\right)^\beta. \tag{4}$$

The parameter ρ_0 is the relative density in a powder situation with point-like contacts between particles. It should be pointed out that Eq. 3 defines a loading surface characterized by a given pair of (ε_{eq}^p, ρ) in the stress space. In the case of isostatic pressing, the yield function is simplified to the following equation with having $\bar{S}_2 = 0$:

$$\sqrt{h}S_1 - k\varepsilon_{eq}^p{}^m = 0. \tag{5}$$

It is worthy to mention that $S_1 = 3P$ in isostatic pressing where P is the applied pressure. And also:

$$\dot{\varepsilon}_{eq}^p = \sqrt{2}\dot{\varepsilon}_1^p. \tag{6}$$

Materials Research Forum LLC
https://doi.org/10.21741/9781644902837-22

The mass conservation implies that:

$$\dot{\rho}^p = 3\rho\dot{\varepsilon}_1^p .$$ (7)

We finally have:

$$\dot{\varepsilon}_1^p = \frac{2hS_1}{2\sqrt{2}mk^2\varepsilon_{eq}^p{}^{2m-1}-3\rho S_1^2\frac{dh}{d\rho}}\hat{S}_1.$$ (8)

As a pair of ρ, ε_{eq}^p is obtained uniquely by plastic mechanisms of deformation in the above formulation, we reshow $\dot{\varepsilon}_1^p$, S_1 and $\hat{S}_1$ as $\dot{\varepsilon}_1^{p*}$, S_1^* and $\hat{S}_1^*$, respectively and rewrite Eq. 8 as the following equation:

$$\dot{\varepsilon}_1^{p*} = \frac{2hS_1^*}{2\sqrt{2}mk^2\varepsilon_{eq}^p{}^{2m-1}-3\rho S_1^{*2}\frac{dh}{d\rho}}\hat{S}_1^* .$$ (9)

Ideal Viscoplasticity of Powder. The intention from ideal viscoplasticity is in fact the permanent deformation resulted from the secondary creep mechanisms of deformation. For this aim, the viscoplastic dissipation potential of the Odqvist's type (ϕ) is used [2]:

$$\emptyset = \frac{A}{n+1}\sigma_{eq}^{n+1} .$$ (10)

Where A and n are two material's parameters depending on temperature. The equivalent stress (σ_{eq}) is selected as the one suggested by Abouaf [3]:

$$\sigma_{eq} = \left(fS_1^2 + \frac{3}{2}c\bar{S}_2^2\right)^{\frac{1}{2}} .$$ (11)

Where f and c are two functions of relative density. In the viscoplastic case, we have:

$$\dot{\varepsilon}_{ij}^{vp} = \frac{\partial\emptyset}{\partial\sigma_{ij}} .$$ (12)

We can therefore obtain the following equation in the case of isostatic pressing:

$$\dot{\varepsilon}_1^{vp} = A(3P)^n f^{\frac{n+1}{2}} .$$ (13)

Function f is considered as the following equation in this study.

$$f = a\left(\frac{1-\rho}{\rho-\rho_0}\right)^b .$$ (14)

Plastic-Viscoplastic Modeling. In a material loading process, one can generally assume two simultaneous deformations:
- An ideal viscoplastic deformation at each stress state,
- A plastic deformation due to the stress increment at each stress state.

Materials Research Forum LLC
https://doi.org/10.21741/9781644902837-22

In the case of isostatic pressing of a powder, the axial viscoplastic strain rate can simply be calculated by Eq. 13. This viscoplastic strain partially contribute to the densification. However, by assumption of ideal viscoplasticity, no material hardening happens by viscoplastic mechanisms of deformation. Concerning the plastic part of deformation, it contributes not only to densification but also to the increase of ε_{eq}^p. Now, imagine a pair of ρ (achieved by both plastic and viscoplastic mechanisms) and ε_{eq}^p. It is evident that since ρ is obtained by both mechanisms, this pair would be achieved at a value of S_1 inferior to the value of S_1^* that is required to obtain the same pair of $(\varepsilon_{eq}^p, \rho)$ by uniquely plastic mechanism (Fig. 2):

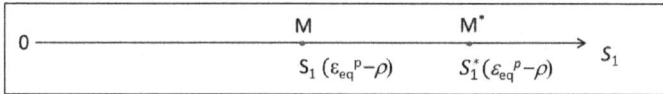

$$
\begin{array}{c}
\text{M} \qquad\qquad \text{M}^* \\
0 \xrightarrow{\hspace{8cm}} S_1 \\
S_1\,(\varepsilon_{eq}{}^p - \rho) \qquad S_1^*(\varepsilon_{eq}{}^p - \rho)
\end{array}
$$

Figure 2. Comparison of S_1 and S_1^* to give a given pair of $(\varepsilon_{eq}^p, \rho)$.

It is therefore necessary to do some logical assumption to calculate $\dot{\varepsilon}_1^p$ at point M in Fig. 2 specified with S_1. As point M^* is the radial projection of M on the loading surface characterized by $(\varepsilon_{eq}^p, \rho)$, we make the following hypothesis:

$\dot{\varepsilon}_1^p$ at point M is proportional to $\dot{\varepsilon}_1^{p*}$ at M^* when the same value is selcted for $\hat{S}_1$ and $\hat{S}_1^*$ ($\hat{S}_1 = \hat{S}_1^*$). The proportionality factor (ω) is considered to be equal to $\frac{S_1}{S_1^*}$. Therefore, regarding Eq. 9, we have:

$$\dot{\varepsilon}_1^p = \omega \frac{2hS_1^*}{2\sqrt{2}mk^2\varepsilon_{eq}^{p\ 2m-1} - 3\rho S_1^{*2}\frac{dh}{d\rho}}\hat{S}_1 \,. \tag{15}$$

Or:

$$\dot{\varepsilon}_1^p = \frac{2hS_1}{2\sqrt{2}mk^2\varepsilon_{eq}^{p\ 2m-1} - 3\rho S_1^{*2}\frac{dh}{d\rho}}\hat{S}_1. \tag{16}$$

Therefore, by a plastic-viscoplastic mechanism of deformation, the viscoplastic and plastic strain rates defined by Eqs. 13 and 16, respectively, happen during isostatic pressing.

Materials and Experiment
Some samples were prepared at Bodycote Inc.. The samples were some cylindrical containers including stainless steel 316L powder. The container was from low carbon steel. The container thickness, internal diameter and internal height were 0.89, 17.3 and 76.2 mm, respectively [4].

The value of ρ_0 was measured equal to 0.6684. The applied isostatic pressing included two steps. First, the sample was heated to 1115°C at a very low pressure. The obtained relative density at this stage is 0.71. Then at 1115°C, different pressure ramps were applied to the sample to achieve a final pressure. Then the heating was cut and the sample was cooled. The relative density attributed to the selected pressure ramp and final pressure was measured. The applied pressure ramp and the final pressure are cited in Table 1. It should be mentioned that each test was repeated three times and a good stability in the results were observed.

Table. 1. Values of pressure ramp and the final pressure achieved in HIP experiments

P (MPa)	2	7	18	24	30
Pressure rate (MPa/min.)	1.8	3.3	1	0.9	1.3

Materials Identification

Despite the presence of container, it is approximated that the densification of stainless steel 316L powder occurs without a major influence of container on the pressure transfer to the powder at 1115°C. Therefore, it is necessary to identify the stainless steel 316L parameters. The following parameters are obtained from literature [4]:

$$lnA = \frac{-44229}{T} + 17.125. \tag{17}$$
$$n = -0.0057T + 9.926. \tag{18}$$
$$m = -0.0001T + 1655. \tag{19}$$
$$k = -0.8886T + 1655. \tag{20}$$
$$a = 0.23 \tag{21}$$
$$b = 1.1 \tag{22}$$

Where T is temperature in Kelvin. It is also necessary to identify function h.

Numerical Simulation

A Fortran program based on Eqs. 13 and 16 was written and run to simulate the powder densification under applying different pressure ramps (Table 1) at 1115°C. The function h was adjusted so that the relative densities at the final pressure, given by experiment and by simulation, were comparable. Therefore, the values of α and β were approximated as 15 and 0.7. The simulation results are shown and compared with the experimental ones in Fig. 3.

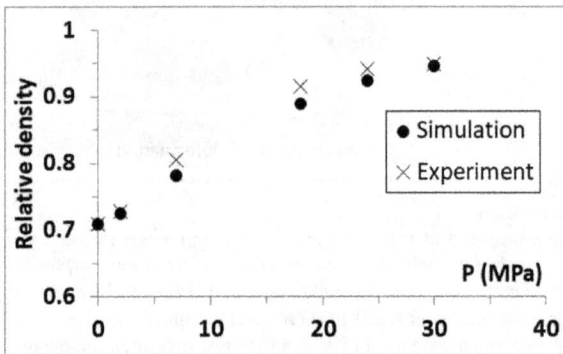

Figure 3. Comparison of plastic-viscoplastic simulation results with the experimental ones.

Conclusion

A plastic-viscoplastic model was introduced for inelastic deformation and thus densification of powders in the HIP process. While relative density is an isotropic hardening parameter for the plastic and viscoplastic behaviors, it is considered that the plastic behavior is also influenced by the equivalent plastic strain as an additional isotropic hardening parameter. With this consideration, the plastic and viscoplastic components of the axial strain in isostatic pressing were derived in this work The model was applied to the HIP of stainless steel 316L powder at 1115°C while different pressure ramps were applied to the powder. The experimental and simulation

Materials Research Forum LLC
https://doi.org/10.21741/9781644902837-22

results were compared. Satisfactory results were obtained. The model must be generalized to the deviatoric stress loading in the next step.

References

[1] R.J. Green. A plasticity theory for porous solids. Int. J. Mech. Sciences, 14 (1972) 215–224. https://doi.org/10.1016/0020-7403(72)90063-X

[2] J. Lemaitre and J. L. Chaboche: 'Mecanique des materiaux solides', 2nd edn; 1996, Paris, Dunod.

[3] M. Abouaf, JL. Chenot, G. Raisson, P. Bauduin, Finite element simulation of hot isostatic pressing of metal powders. Int J Numer Methods Eng 25 (1988), 191–212. https://doi.org/10.1002/nme.1620250116

[4] G. Aryanpour, S. Mashl, V. Warke, Elastoplastic–viscoplastic modelling of metal powder compaction: application to hot isostatic pressing, Powder Metallurgy, 56 (2013). https://doi.org/10.1179/1743290112Y.0000000027

Hot Isostatic Pressing - HIP'22   Materials Research Forum LLC
Materials Research Proceedings 38 (2023) 166-171   https://doi.org/10.21741/9781644902837-23

HIP Modeling and Design of Large Complex Shape Parts Close to the Size of the HIP Furnace Accounting Capsules Manufacturing Technology and their Movement Inside it During the Cycle

Dmitry Seliverstov[1,a], Evgeny Khomyakov[1,b], Alex Bissikalov[1,c], Victor Samarov[1,d*]

[1]Synertech PM, Garden Grove, CA, USA

[a]Dmitry@synertechpm.com, [b]Evgeny@synertechpm.com, [c]Alex@synertechpm.com, [d]Victor@synertechpm.com

Keywords: HIP Modeling, Large Complex Shape Parts

Abstract. The size of the parts needed to be HIPed for critical applications, especially in nuclear, is growing and very often the HIP capsules become so large that they take not only the entire diameter of the HIP furnace, but almost all its volume. Often, these parts weigh thousands of pounds, do not have axial symmetry and cannot stand by themselves in the furnace requiring complex shape supporting frames. For such parts additional problems need to be solved in the HIP tooling design and HIP process itself due to: (1) difficulties in maintaining dimensional accuracy, (2) non-uniform temperature field inside the HIP during heating and ramping the pressure, (3) different mechanical properties in HIP capsule elements and (4) HIP capsule movement during deformation inside the supporting frame. HIP modeling allows to approach these problems. Several examples of the solutions allowing HIPing of complex shape non-symmetrical parts up to 70" in size are analyzed.

Introduction

The two major tasks for HIP of large parts are:

1. Densification to 100% and formation of properties. This is well understood, though requires substantial efforts in enhancing powder manufacturing, powder storage and handling, and pre-HIP processing.

2. Shape control to NNS during densification. This is provided through HIP process FEM modelling and capsule design and "freezing" of the powder tap density and PSD.

However, exceptional uniformity of temperature (10-25F) and isostatic pressure at the end of the HIP cycle is combined with a large non-uniformity (100-200F) at the ramp stage of it. Also, no (or very little) densification happens at the end of the hold temperature when the material is practically 100% dense, but initial deformation of the large cans with powder occurs in a non-uniform temperature field.

HIP of NNS Large Parts

The size of the NNS PM HIPed parts needed for critical applications is growing and very often the HIP capsules become so large that they take not only the entire diameter of the HIP furnace, but almost all its entire height. Very often, these parts are heavy, weigh thousands of pounds, do not have an axial symmetry and cannot stand by themselves in the furnace, requiring complex shape supporting frames (Fig. 1). This leads to the non-uniformity of the powder temperature at the most important stage of the HIP cycle when densification of powder and shape changes mostly occur, and deformation pattern is defined.

Figure 1. Large HIP capsules taking the entire (a) diameter and (b) volume of the HIP furnace.

Other stochastic technological factors and instability of the process parameters and material properties that can increase random geometrical scatter during HIP are:

- Mechanical properties of the steel sheet used for fabrication of capsule elements (chemical analysis, yield strength, thermo- mechanical processing, thickness).
- Powder variation in tap density / particle size distribution (PSD).
- Welding process, especially in hand welding (stiffness, distortions due to thermal stresses).
- Variations in the HIP cycle trajectory.
- The value of the powder tap density and its variations within the capsule. Need to "freeze" the parameters of atomization process.
- Location of a capsule in a HIP furnace may be important due to non-uniform radiation heating followed by preferential deformation of the areas getting hot first.

An important role of HIP Modeling is to help to determine the sensitivity of the final geometry to these parameters through multiple virtual HIP iterations and to provide the design of the capsule the least sensitive to these variations.

HIP tooling design is the core operation in the development of the NNS parts. Since HIP takes place in a freely plastically deformed tool that is not well supported by the densifying powder at the initial stages of HIP, when the yield strength of the capsule material is still relatively high, the balance of the stiffness of the capsule elements is essential. While the axial deformation (for the parts with a kind of axial symmetry) is like stable uni-axial densification, the radial deformation can lead to macro-bending and other macro-distortions, if the capsule is not properly balanced. At the later stages of HIP, when the powder material is partially densified and behaves as a compressible rigid plastic media, the deformation pattern becomes much more uniform, but the distortions introduced at the beginning of the process cannot be compensated and will progress. Sometimes, as will be shown below, the specific conditions of the capsule manufacturing require introduction of the gradient properties of the capsule material that are essential for the HIP deformation of large NNN parts. Therefore, the behavior of the HIP tooling at the beginning of the process has the greatest influence on the final shape and dimensions of the part. Any macro-non uniformity of the initial capsule is inherited further, can evolve and progress during HIP. Also, a

Hot Isostatic Pressing - HIP'22   Materials Research Forum LLC
Materials Research Proceedings 38 (2023) 166-171   https://doi.org/10.21741/9781644902837-23

capsule design must provide full and uniform filling with powder leaving no "pockets" or area of lower density.

Three Development Stories for HIP of Very Large Parts of SMR
1. 44% Top Head Development - Impact of the Capsule Manufacturing Technology.
The 44% Top Head (Fig. 2) is 51" Ø which is the max size that can fit in the existing largest US HIP (62" Ø) accounting the shrinkage of capsule with A 508 powder.

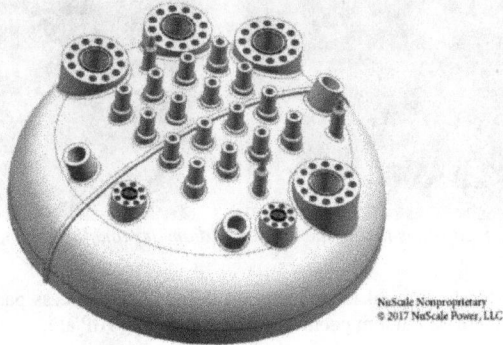

NuScale Nonproprietary
© 2017 NuScale Power, LLC

Figure 2. A508 Top Head of a SMR with integral nozzles

HIP modeling and design of the HIP tooling were based on the developed rheological properties for the capsule steel and A508 powder (Fig. 3).

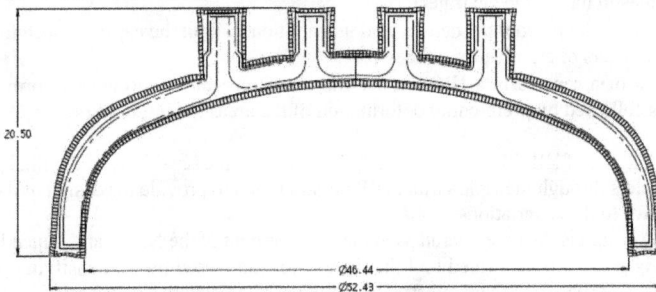

Result of Simulation for Mild Steel Can: Cross-section 2 with Central Bosses

Figure 3. HIP modeling for the 44% Top Head providing a machinable blank.

Capsule elements were manufactured through several operations: die forming, spinning of the flanges, laser cutting, welding. The welded capsule can be seen in Fig. 4.

Materials Research Forum LLC
https://doi.org/10.21741/9781644902837-23

Figure 4. Welded capsule for 44% Top Head

As a result of HIP, the material was fully consolidated, but the periphery of the HIP can has not shrunk to the desired values (green) (Fig. 5). It has turned out that during cold spinning the material of the capsule flange has acquired substantial strain hardening that had not been accounted during modeling. As result, the flange had higher strength than the rest of the capsule, that had impacted the shrinkage. HIP modeling accounting the Strain Hardening of the Capsule Flange has confirmed the received dimensions. The new capsule material properties have been introduced into the data base for this type of capsules.

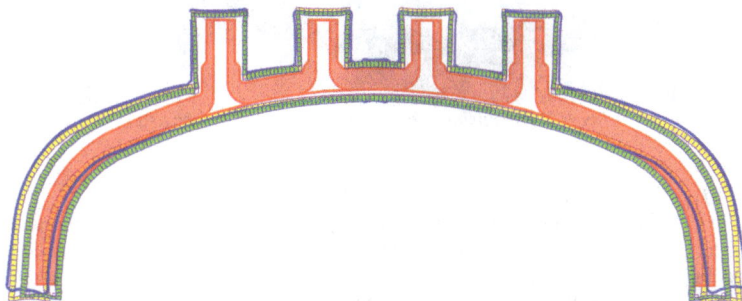

Figure 5. Dimensional analysis of modeling and actual shrinkage

2. 67% ½ Lower Head - HIP of a non-symmetrical capsule in a frame.
The 89" Ø 67% lower head could be manufactured only in two halves in vertical position in the supporting frame (Fig. 1b). HIP modeling was done accounting capsule deformation in the frame with local friction and sliding (Fig. 6a). The manufactured lower head can be seen in Fig. 6b.

Hot Isostatic Pressing - HIP'22
Materials Research Proceedings 38 (2023) 166-171

Materials Research Forum LLC
https://doi.org/10.21741/9781644902837-23

(a)

(b)

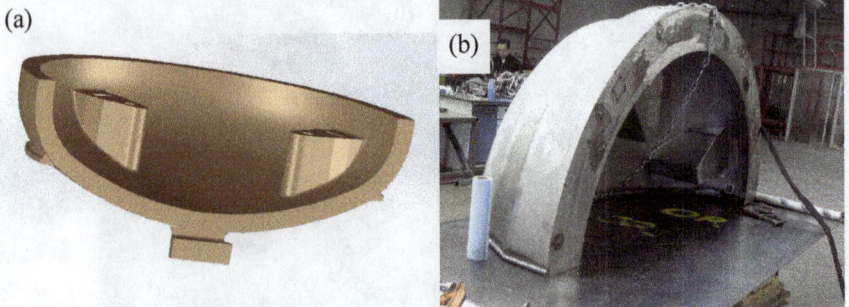

Figure 6. (a) A508 lower head of the SMR and (b) HIP capsule for the lower head

Laser scanning of the HIPed lower head has confirmed the results of modeling (Fig. 7).

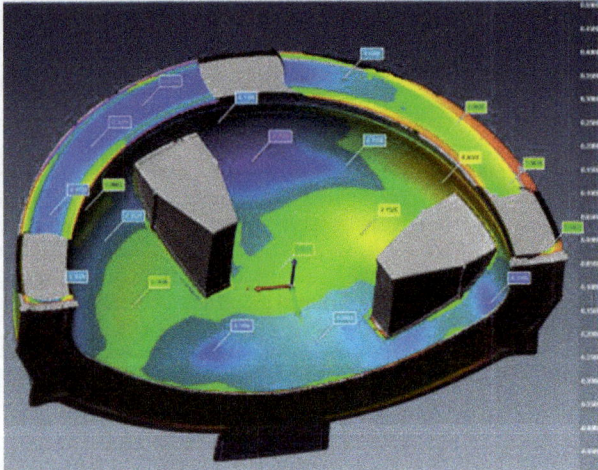

Figure 7. Dimensional analysis of the HIPed lower head

3. 67% ½ Top Head Development.
The 80" Ø 67% top head could be manufactured only in two halves also using a supporting frame. The general deformation of the two capsule shells and location of nozzles were defined through HIP modeling. An image of the head following laser cutting can be seen in Fig. 8. The ½ top head after powder filling (Fig. 9a) and in the shape support frame (Fig. 9b).

Figure 8. CNC laser cutting operations.

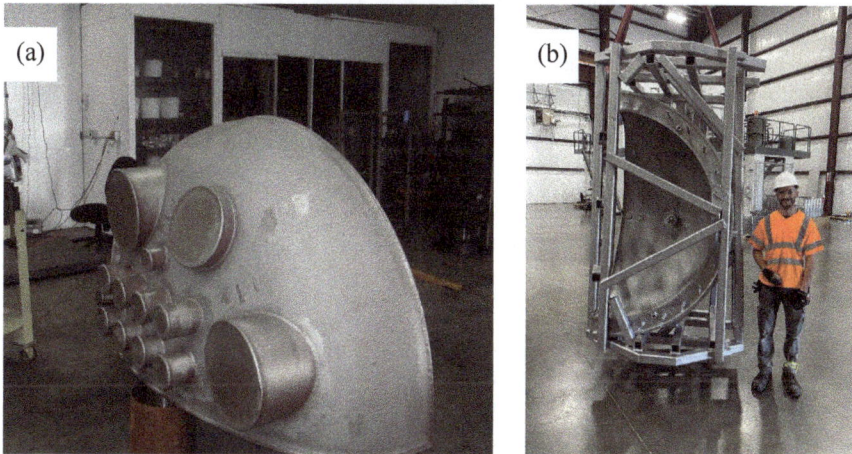

Figure 9. HIPed 1/2 top head (a) after powder filling and (b) in the support structure

Conclusions

Problems to be analyzed and solved for HIP of large parts close to the size of the HIP furnace are:

1. Non uniformity of the temperature field during the HIP ramp leading to non-uniform deformation of capsules with powder.
2. Preferential deformation of the capsules or capsule elements affected by radiation from the heaters.
3. Capsule manufacturing processes that have a strong impact on the deformation during HIP.
4. Large non-symmetrical capsules need frames to position them in the furnace and guide their deformation during HIP.
5. Efficient design and modeling tools, not just modeling art are needed.
6. With all these factors accounted and problems solved it becomes possible to manufacture large NNS parts from the first shot.

Materials Research Forum LLC
https://doi.org/10.21741/9781644902837-24

Effect of Experimental Determination Process on Shear Stress Coefficient of Green Equation Describing HIP

Gerard Raisson[1,a,*], Vassily Goloveshkin[2,b], Evgeny Khomyakov[3,c], Victor Samarov[4,d]

[1]Consultant, Cournon D'Auvergne, France

[2]IAM RAS, RTU MIREA, Moscow, Russia

[3]Synertech PM Inc., Garden Grove, CA, USA

[4]LNT PM Inc., Garden Grove, CA, USA

[a]gerard.raisson@gmail.com, [b]nikshevolog@yandex.ru, [c]evgeny@synertechpm.com, [d]victor@synertechpm.com

Keywords: HIP Modeling, HIP Tooling Design

Abstract. Despite the uniformity of the boundary conditions on the surface of the HIP capsule during the HIP Cycle (Pressure, Temperature) and full uniformity of stresses inside the powder material at the end of the cycle (isotropic tensor), the compressive deformation of the capsules with powder has a substantial shear component that can be described as distortion or non-uniformity. Accounting the values of shear deformation is the subject of HIP modeling and HIP tooling design. To determine the functions that adequately describe deformation of capsules with the given powder material during HIP, it is important to carry out experiments so that the values of the deformation tensor are close to those in the powder during a real HIP process. The adequacy of 3 deformation conditions for a combination TA6-4 powder and 304L capsule was evaluated for the database generation: HIP in capsules with different stiffness (deformation close to flat); Uni-directional upsetting of porous samples; Upsetting in a rigid die (uni-direction deformation). After homogenization of full dense material and capsule material, coefficients of Green's equation have been calculated for TA6-4 powder in a 304L capsule. Unidirectional densification and HIP dilatometer give similar results when upsetting results are apart. Analysis of results shows that, in actual conditions of HIP cycle, only a small portion of ellipsoidal plasticity surface given by Green's equation is used. Unidirectional pressing and HIP dilatometer condition are on the actual portion of ellipse. Upsetting of porous material is out of the useful zone and seems not convenient to determine shear stress coefficients.

Introduction

To determine the functions that adequately describe deformation of capsules with the given powder material during HIP, it is important to carry out experiments so that the values of the deformation tensor are close to those in the powder during a real HIP process.

Analysis of results shows that, in actual conditions of HIP cycle, only a small portion of ellipsoidal plasticity surface given by Green's equation is used for the actual stress-strain conditions:

1. Deformation of porous media and plasticity surface
2. At every elementary increment of temperature and pressure the powder material is moving to a new plasticity surface
3. Plasticity surface is presented as a set of changing with ellipsoidal surfaces in the space of the main stresses (1-1-1) and deformation of the material
4. Occurs in some vicinity of the axis, on the "cap" of the ellipsoid

Hot Isostatic Pressing - HIP'22   Materials Research Forum LLC
Materials Research Proceedings 38 (2023) 172-176   https://doi.org/10.21741/9781644902837-24

To describe the behavior of the powder material the elliptical Green's plasticity criterion was used:

$$\frac{\sigma^2}{f_2^2} + \frac{s^2}{f_1^2} = Y^2 . \tag{1}$$

where: $\sigma = \frac{1}{3}\left(\sigma_r + \sigma_\varphi + \sigma_z\right)$ - the average stress (σ_{ij} - the corresponding components of the stress tensor);

$$s^2 = \frac{3}{2}\left[\frac{\left(2\sigma_r - \sigma_\varphi - \sigma_z\right)^2}{9} + \frac{\left(2\sigma_\varphi - \sigma_r - \sigma_z\right)^2}{9} + \frac{\left(2\sigma_z - \sigma_r - \sigma_\varphi\right)^2}{9} + 2\sigma_{r\varphi}^2 + 2\sigma_{rz}^2 + 2\sigma_{\varphi z}^2\right]$$

intensivity of the stress tensor deviator; f_1, f_2 - functions of powder density ρ defined from experiments; Y - yield strength of the monolithic material.

"Classical" Experiments to define the functions f_1, f_2.

First experiment-HIP in a thin-walled capsule.

If we ignore the influence of the capsule walls, and $\sigma_r = \sigma_\varphi = \sigma_z = -P$

Then from relation (1), we get: $\frac{P^2}{f_2^2} = Y^2 \Rightarrow f_2^2 = \frac{P^2}{Y^2} \Rightarrow f_2 = \frac{P}{Y}$.

Second experiment- upsetting of a cylindrical porous sample; $\sigma_r = \sigma_\varphi = 0, \sigma_z = -T$

From relation (1), we get: $\frac{T^2}{9f_2^2} + \frac{T^2}{f_1^2} = Y^2$.

As far as $\frac{1}{f_2^2} = \frac{Y^2}{P^2}$, then $\frac{1}{f_1^2} = \frac{Y^2}{T^2}\left(1 - \frac{T^2}{9P^2}\right)$

The vector of stresses in the principal axes is $\{0;0;-T\}$. The hydrostatic axis has a direction $\{-1;-1;-1\}$ and the angle between them is ψ. Then $\cos\psi = 1/\sqrt{3} = 0.577$. $\psi \approx 0.95 \approx 54^0$.

This means that free upsetting of porous samples occurs at the angle of 54 degrees from the hydrostatic axis.

Now let us explore the real stress conditions of the HIP process. For the strain rates during deformation of power material we have:

$$\varepsilon_{ij} = \omega\frac{\partial\Phi}{\partial\sigma_{ij}} . \tag{2}$$

where: - ε_{ij} - the corresponding components of the strain rate tensor; and $\Phi\left(\sigma_{ij}\right) = 0$ equation for the flow surface.

Let us consider that $\varepsilon_r, \varepsilon_\varphi, \varepsilon_z$ - main components of the strain rate tensor. In the majority of the HIP processes (can geometries) these components are compressive,

Let us designate $\dfrac{\varepsilon_\varphi}{\varepsilon_r} = \alpha$ $\dfrac{\varepsilon_z}{\varepsilon_r} = \beta$ and consider $0 \le \beta \le \alpha \le 1$. Using (1, 2), we can define the important value of the ANGLE between the "vector of stresses" and hydrostatic axis.

$$\cos\Psi = \frac{1}{\sqrt{3}} \times$$

$$\times \frac{\left\{27 f_2^2\right\}}{\sqrt{\left[9 f_2^2 + 2 f_1^2 \dfrac{(2-\alpha-\beta)}{(1+\alpha+\beta)}\right]^2 + \left[9 f_2^2 + 2 f_1^2 \dfrac{(-1+2\alpha-\beta)}{(1+\alpha+\beta)}\right]^2 + \left[9 f_2^2 + 2 f_1^2 \dfrac{(-1-\alpha+2\beta)}{(1+\alpha+\beta)}\right]^2}} \tag{3}$$

Analyzing this relationship, it is possible to show that at $0 \le \beta \le \alpha \le 1$:

$$\cos\Psi \ge \frac{\left\{9 f_2^2\right\}}{\sqrt{81 f_2^4 + 8 f_1^4}} \tag{4}$$

In reality the value of f_1 is usually less than f_2. Therefore

$$\cos\Psi \ge \frac{9}{\sqrt{89}} \approx 0.954. \text{ Then } \psi < 0.31 \approx 18^0. \tag{5}$$

This means that in the real HIP conditions the angle ψ is less than 18 degrees.

In order to be in the real stress condition compatible with HIP, the second experiment for the definition of f_1 and f_2 is better to conduct as uni-directional deformation of powder in a rigid die.

Considering that $\sigma_z = -F$ - is a known value and. $\varepsilon_r = \varepsilon_\varphi = 0$

Then:

$$\sigma_z = \frac{1}{\omega}\left[\left(\frac{9 f_2^2 - 2 f_1^2}{18}\right)\varepsilon_r + \left(\frac{9 f_2^2 - 2 f_1^2}{18}\right)\varepsilon_\varphi + \left(\frac{9 f_2^2 + 4 f_1^2}{18}\right)\varepsilon_z\right] \tag{6}$$

$$\frac{1}{\omega} = \frac{6Y}{\sqrt{\left(9 f_2^2 - 2 f_1^2\right)\left(\varepsilon_r + \varepsilon_\varphi + \varepsilon_z\right)^2 + 6 f_1^2\left(\varepsilon_r^2 + \varepsilon_\varphi^2 + \varepsilon_z^2 + 2\varepsilon_{r\varphi}^2 + 2\varepsilon_{\varphi z}^2 + 2\varepsilon_{rz}^2\right)}} \tag{7}$$

To

$$\sigma_z = \frac{6Y}{\sqrt{\left(9f_2^2 - 2f_1^2\right)\left(\varepsilon_r + \varepsilon_\varphi + \varepsilon_z\right)^2 + 6f_1^2\left(\varepsilon_r^2 + \varepsilon_\varphi^2 + \varepsilon_z^2 + 2\varepsilon_{r\varphi}^2 + 2\varepsilon_{\varphi z}^2 + 2\varepsilon_{rz}^2\right)}} \times$$
$$\times \left[\left(\frac{9f_2^2 - 2f_1^2}{18}\right)\varepsilon_r + \left(\frac{9f_2^2 - 2f_1^2}{18}\right)\varepsilon_\varphi + \left(\frac{9f_2^2 + 4f_1^2}{18}\right)\varepsilon_z\right] \tag{8}$$

$$\sigma_z = \frac{6Y}{\sqrt{\left(9f_2^2 - 2f_1^2\right)\varepsilon_z^2 + 6f_1^2\varepsilon_z^2}}\left[\left(\frac{9f_2^2 + 4f_1^2}{18}\right)\varepsilon_z\right] \tag{9}$$

$$\sigma_z = -\frac{Y\left(9f_2^2 + 4f_1^2\right)}{3\sqrt{9f_2^2 + 4f_1^2}} \tag{10}$$

$$F = \frac{Y}{3}\sqrt{9f_2^2 + 4f_1^2} \tag{11}$$

$$\frac{9F^2}{Y^2} = 9f_2^2 + 4f_1^2 \tag{12}$$

As $f_2^2 = \dfrac{P^2}{Y^2}$, then

$$\frac{9F^2}{Y^2} = 9\frac{P^2}{Y^2} + 4f_1^2 \Rightarrow 9\left(\frac{F^2}{Y^2} - \frac{P^2}{Y^2}\right) = 4f_1^2 \tag{13}$$

$$f_1^2 = \frac{9}{4}\frac{F^2 - P^2}{Y^2} \Rightarrow f_1 = \frac{3}{2Y}\sqrt{F^2 - P^2} \tag{14}$$

We can then build the following Table of Angles on the Plasticity Surface validating the experimental techniques for defining the f_1 and f_2 plasticity functions.

Hot Isostatic Pressing - HIP'22
Materials Research Proceedings 38 (2023) 172-176

Materials Research Forum LLC
https://doi.org/10.21741/9781644902837-24

Table 1. Angles on the Plasticity Surface for Different Processes

N	Condition	Angles on the Plasticity Surface for different processes, degrees	Comments
1	Hydrostatic conditions (very thin or very soft can material)	0	1st "classical" experiment, can work for high strength PM alloys, defines the function f_2
2	Free upsetting of porous samples obtained in the interrupted cycles	54	2nd "Classical" experiment" to define f_1, but the angle ψ is far from HIP working area
3	Uni-directional compressing	18	Difficult to accomplish, needs a rigid die
4	"Flat" deformation, $\varepsilon_z = 0$	9.2	Valid for HIP of a long cylinder
5	Maximal angle during HIP	18	Can be reached when the capsule is very rigid in two directions and the deformation occurs mainly in one direction

Conclusions

Despite the uniformity of the boundary conditions on the surface of the HIP capsule during HIP and full uniformity of stresses inside the powder material at the end of the cycle, the compressive deformation of the capsules with powder has a substantial shear component. Accounting these values of shear deformation is the subject of HIP modeling and HIP tooling design.

To determine the functions that adequately describe deformation of capsules with the given powder material during HIP, it is important to carry out experiments so that the values of the stress tensor are close to those in the powder during a real HIP process.

Theoretical analysis of results shows that, in actual conditions of HIP cycle, only a small portion of ellipsoidal plasticity surface in the space of main stresses given by Green's equation (the "cap") is used.

Unidirectional pressing, flat deformation and HIP dilatometer condition are on the actual portion of ellipsoidal surface. Upsetting of porous material is out of the useful zone to determine shear stress coefficients.

Hot Isostatic Pressing - HIP'22   Materials Research Forum LLC
Materials Research Proceedings 38 (2023) 177-183   https://doi.org/10.21741/9781644902837-25

Theoretical Evaluation of Capsule Material Strain Hardening on the Deformation of Long Cylindrical Blanks During HIP Process

Andrey Bochkov[1,a], Yury Kozyrev[1,b], Anton Ponomarev[2,c], Gerard Raisson[3,c,*]

[1]RTU MIREA, Moscow, Russia

[2]IAM RAS, RTU MIREA, Moscow, Russia

[3]Consultant, Cournon D'Auvergne, France

[a]andrey.bochkov@gmail.com, [b]gmile88@mail.ru, [c]avpon@yandex.ru,
[d]gerard.raisson@gmail.com

Keywords: HIP Process, Powder Material, Non-Uniform Temperature Field, Yield Criterion of Green

Abstract. The paper presents the results of the theoretical study of the effect of strain rate hardening (viscosity) of the capsule material on the process of hot isostatic pressing of a long cylindrical can with powder (so that the influence of lids can be neglected). Relations are obtained that make it possible to evaluate the qualitative estimations of the processes taking place at different stages of the HIP process. Calculations are made for the capsule shrinkage values.

Introduction

Powder Metallurgy and HIP have been established as a promising method of manufacturing large complex shape parts operating at harsh environment and loads. For such parts made of expensive materials high dimensional precision in production is needed. This is especially true for small-scale production of large-sized products, when the cost of trial samples can be very high.

The HIP process that is conducted at high temperature and pressure causes substantial (30-35%) volumetric changes of a capsule filled with powder material and, if necessary, with inserts embedded in certain places, which are subsequently removed from the finished product. During such HIP process, this capsule takes large non-uniform linear deformations, up to 20%. This deformation occurs under changing temperature and with non-uniform temperature field inside the capsule and the capsule itself exhibits plastic and viscous deformations. It is the complexity of modeling the process of distortion of the workpiece that makes it difficult to obtain the product of the required geometry from the "first shot".

Therefore adequate modeling of the geometrical changes during HIP of the resulting product becomes a mandatory technological tool, which can be carried out by mathematical modeling methods, some aspects of which are considered in [1, 2]. In the most general form, the formulation of the problem of mathematical modeling of the process of HIP can be formulated as follows: it is required to design the capsule in such a way that the final shape and dimensions of the powder part obtained after removal of the capsule satisfies the required geometry. Note that due to the specifics of the use of such products, these requirements are quite stringent.

For a qualitative understanding of various aspects of the process of HIP and further comparison of the experimental data with the results of numerical experiments, a theoretical study of the effect of various factors on the process of HIP is necessary. In this paper, we investigate the effect of viscosity of the capsule material on the process of HIP of a long cylindrical billet, which is one of the typical problems to be solved in the production.

Mathematical Statement of the Problem of Modeling of the HIP Process

The powder material is considered as a single plastically compressible medium. The problem of HIPing a plastically compressible material in a capsule made of a non-compressible material with

Hot Isostatic Pressing - HIP'22   Materials Research Forum LLC
Materials Research Proceedings 38 (2023) 177-183   https://doi.org/10.21741/9781644902837-25

strain hardening under conditions of an inhomogeneous nonstationary temperature field is considered.

The general mathematical statement of the problem of modeling the process of HIP includes the following:

Equilibrium equation:

$$div\overline{\sigma} = 0.$$ (1)

where $\overline{\sigma}$ – is the stress tensor.

The equation of the yield surface in the general case is given in the form:

$$\Phi\left(\sigma_{ij}\right) = 0.$$ (2)

The relationship between the stress tensor σ_{ij} and the strain rate tensor ε_{ij} is determined by the associated flow law:

$$\varepsilon_{ij} = \omega \frac{\partial \Phi}{\partial \sigma_{ij}}.$$ (3)

where ε_{ij} – is the strain rate tensor.

To describe the behavior of the capsule material and the embedded inserts, the ideal plasticity condition and incompressibility condition are used:

$$S^2 = T^2.$$ (4)

$$div\overline{u} = 0.$$ (5)

where T – is the yield strength, S^2 – the intensity of the deviator of the stress tensor.

To determine the density, the continuity equation is used:

$$\frac{d\rho}{dt} + \rho div\overline{u} = 0.$$ (6)

where $\overline{u}$ – is the displacement speed.

The problem is considered in a quasi-static formulation, the process of deformation is considered to be rather slow, therefore, the acceleration in the equilibrium equations can be neglected. The current formulation of this does not take into account the temperature expansion of the materials.

HIP of a long Cylinder

Let us consider the process of hot isostatic pressing of a cylinder, assuming that its height greatly exceeds the radius. The influence of the capsule top and bottom caps can be then neglected.

Suppose that the axis Z is directed along the axis of the cylinder. The problem is considered in an axisymmetric formulation in a cylindrical coordinate system. We will assume that the region

$0 < r < R, 0 < z < H$ is occupied by a powder material, and the region $R < r < R + h, 0 < z < H$ –
by a capsule.

To describe the mechanical properties of the powder material, we will use the Green's yield criterion [3,4]:

$$\frac{\sigma^2}{f_2^2} + \frac{S^2}{f_1^2} = Y^2 .$$

(7)

where $\sigma = \frac{1}{3}\left(\sigma_r + \sigma_\varphi + \sigma_z\right)$ – the average stress; $S^2 = \frac{3}{2}\left(\sigma_{ij} - \sigma\delta_{ij}\right)\left(\sigma_{ij} - \sigma\delta_{ij}\right)$ –intensity of stress tensor deviator; Y – yield stress of a fully dense powder material; f_1 and f_2 – experimental functions of relative density ρ.

We assume that the rate of deformation along the Z symmetry axis is constant throughout the volume of the powder material and the capsule. In view of the incompressibility condition (5), in the capsule material, and the condition of continuity of the radial velocity $r = R$, we obtain the strain rates are:

1) at $0 < r < R, 0 < z < H$ (in the powder material)

$$\varepsilon_z = -\varepsilon , \; \varepsilon_r = -A , \; \varepsilon_\varphi = -A .$$

(8)

2) at $R < r < R + h, 0 < z < H$ (in the capsule material)

$$\varepsilon_z = -\varepsilon , \; \varepsilon_r = \frac{1}{2}\varepsilon + \frac{1}{2}\varepsilon\frac{R^2}{r^2} + A\frac{R^2}{r^2} , \; \varepsilon_\varphi = \frac{1}{2}\varepsilon - \frac{1}{2}\varepsilon\frac{R^2}{r^2} - A\frac{R^2}{r^2} .$$

(9)

In this case, the total power of the internal forces in the powder material is:

$$W_p = \pi R^2 H \frac{Y}{3} \sqrt{\left(9f_2^2 - 2f_1^2\right)\left(2A + \varepsilon\right)^2 + 6f_1^2\left(2A^2 + \varepsilon^2\right)} .$$

(10)

In addition, the total power of the internal forces in the capsule:

$$W_c = 2\pi HT \sqrt{\frac{2}{3}} \int_R^{R+h} \sqrt{\frac{3}{2}\varepsilon^2 + 2\left(A + \frac{1}{2}\varepsilon\right)^2 \frac{R^4}{r^4}} \, r dr .$$

(11)

The external HIP pressure is assumed to be equal P. In this case, the total power of the external forces at the surface of the capsule can be expressed as follows:

$$N = P\pi HR^2\left(\varepsilon + 2A\right).$$

(12)

Assuming $\left(\varepsilon + 2A\right) = 1$ and denoting H_0 the value H and R_0 the value R for $\rho = \rho_0$, we get:

$$H(\rho) = H_0 \exp\left\{-\int_{\rho_0}^{\rho} \frac{Y}{Y \cdot \dfrac{3f_1^2}{\sqrt{\left(9f_2^2 - 2f_1^2\right)}} + 2\sqrt{3}\dfrac{h}{R}T} \cdot \frac{f_1^2}{\sqrt{\left(9f_2^2 - 2f_1^2\right)}} \frac{d\rho}{\rho}\right\}. \tag{13}$$

The value $R(\rho)$ is given by:

$$R^2(\rho) = R_0^2 \frac{\rho_0}{\rho} \cdot \frac{H_0}{H}. \tag{14}$$

Note that the final geometry depends significantly on the ratio of the yield strength of the powder material and the yield point of the capsule material. In actual processes with relatively fast ramp of temperature and pressure during HIP, the yield strength of the capsule depends essentially on the rate of deformation. Let us examine the influence of this factor on a specific example.

Let the pressure grow with time as $P = P_0 t$. Suppose that the yield strength of the capsule material is dependent on the deformation rate and has the form:

$$T = B\left(\frac{3\sqrt{2}}{2}\right)^{\alpha} \left(\varepsilon^2 + \varepsilon A + A^2\right)^{\frac{\alpha}{2}} = B\left(\frac{3\sqrt{2}}{2}\right)^{\alpha} A^{\alpha}\left(x^2 + x + 1\right)^{\frac{\alpha}{2}}. \tag{15}$$

where $x = \dfrac{\varepsilon}{A} \leq 1$.

Considering dimensionless time τ according to relation $t = \dfrac{Y}{P_0}\tau$ and dimensionless $\overline{A}$ according to $A = \dfrac{P_0}{Y}\overline{A}$ (do not write line further), we get system of equations determine the solution of the entire Task – functions $A(\tau), \rho(\tau), x(\tau)$.

$$\frac{1}{3}\frac{\left[9f_2^2(2+x) + 2f_1^2(1-x)\right]}{\sqrt{9f_2^2(2+x)^2 + 2f_1^2(1-x)^2}} + \\
+ \left[\frac{2f_1^2}{3\gamma}\frac{(1-x)(x+2)}{x\sqrt{\left(1+x+x^2\right)}}\frac{1}{\sqrt{9f_2^2(2+x)^2 + 2f_1^2(1-x)^2}}\right] = \tau \tag{16}$$

where $\gamma = \dfrac{B}{Y}\left(\dfrac{3\sqrt{2}}{2}\dfrac{P_0}{Y}\right)^{\alpha}$ characterizes the pressure loading rate and α determines the relation of the capsule yield strength from the strain rate (strain hardening).

Materials Research Forum LLC
https://doi.org/10.21741/9781644902837-25

The equation for the powder density:

$$\frac{d\rho}{d\tau} = \rho A(2+x).$$ (17)

The reduction of the radius is determined from the relation $\dfrac{dR}{d\tau} = -AR$, then:

$$R = R_0 \exp\left(-\int_0^\tau A(\tau)d\tau\right).$$ (18)

where R_0 – the initial radius.

From the mass conservation law for powder volume we get:

$$\pi\rho_0 R_0^2 H_0 = \pi\rho R^2 H$$

Accounting (18) it is possible to write:

$$H = H_0 \frac{\rho_0}{\rho} \exp\left(2\int_0^\tau A(\tau)d\tau\right).$$ (19)

For the non-compressible capsule material we can write:

$$2\pi R_0 h_0 H_0 = 2\pi RhH$$

where h_0 – the initial capsule wall thickness.

Using (18, 19), we get:

$$h = h_0 \frac{\rho}{\rho_0} \exp\left(-\int_0^\tau A(\tau)d\tau\right).$$ (20)

According to (17-20), we get:

$$A = \frac{1}{\sqrt{(1+x+x^2)}} \left[\frac{2f_1^2}{\gamma\sqrt{3}} \cdot \frac{R_0}{h_0} \cdot \frac{\rho_0}{\rho} \cdot \frac{1-x}{x} \cdot \frac{1}{\sqrt{9f_2^2(2+x)^2 + 2f_1^2(1-x)^2}}\right]^{\frac{1}{\alpha}}.$$ (21)

The system of equations of (16-21) with the corresponding initial conditions defines the final solution of the Task.

For the qualitative analysis, let us consider functions $f_1(\rho)$ и $f_2(\rho)$ as follows:

$$f_1^2(\rho) = \frac{3}{2+\rho^2} \cdot \left(\frac{\rho-\rho_p}{1-\rho_p}\right)^m , \qquad f_2^2(\rho) = \frac{1}{3(1-\rho^2)} \cdot \left(\frac{\rho-\rho_p}{1-\rho_p}\right)^m$$

Below we can see in Table 1 the shrinkage values – the ratios of the initial and final dimension of the HIPed capsule as a function of the material and process parameters: R_f – is the final radius value at full densification, H_f – is the final height value. $e_r = \dfrac{R_0}{R_f}$, $e_z = \dfrac{H_0}{H_f}$

Calculations have been done at $m = 2$, $\dfrac{h_0}{R_0} = 0,025$.

Table 1. Radial and axial shrinkage values.

№	γ	α	e_r	e_z	$\dfrac{e_z}{e_f}$
1		0.1	1.199	1.043	0.870
2	10	0.2	1.192	1.055	0.885
3		0.3	1.189	1.061	0.892
4		0.1	1.184	1.071	0.904
5	5	0.2	1.180	1.078	0.914
6		0.3	1.176	1.084	0.922
7		0.1	1.167	1.102	0.945
8	2	0.2	1.164	1.107	0.951
9		0.3	1.161	1.112	0.957
10		0.1	1.157	1.120	0.967
11	1	0.2	1.156	1.123	0.972
12		0.3	1.154	1.127	0.976
13		0.1	1.214	1.017	0.838
14	15	0.2	1.201	1.041	0.866
15		0.3	1.196	1.048	0.876
16		0.1	1.211	1.023	0.845
17	20	0.2	1.203	1.037	0.863
18		0.3	1.200	1.042	0.868
19		0.1	1.220	1.008	0.827
20	25	0.2	1.206	1.032	0.856
21		0.3	1.203	1.036	0.862

Conclusions

Analytic relationships have been obtained within the framework of the analysis of the process of the HIP of a long cylindrical billet, making it possible to evaluate the qualitative picture of the processes occurring at different stages of deformation of the cylinder. The effect of the viscous properties (strain hardening) of the capsule material on the final geometry of the product was studied. Calculations have been made for the ratio of the initial cylinder sizes to the final ones (HIP shrinkages), as a function of the process and material parameters.

References

[1] Anokhina A.V., Goloveshkin V.A., Samarov V.N., Seliverstov D.G., Raisson G. Matematicheskaya model' rascheta protsessa goryachego izostaticheskogo pressovaniya detaley slozhnoy formy pri nalichii periodicheskoy struktury zakladnykh elementov. Mekhanika kompozitsionnykh materialov i konstruktsiy. [Mathematical model for calculating the process of

Materials Research Forum LLC
https://doi.org/10.21741/9781644902837-25

hot isostatic pressing of details of complex shape in the presence of a periodic structure of embedded elements. Mechanics of composite materials and structures.] v. 8 № 2. p. 245-254.

[2] A.V. Bochkov, V.A. Goloveshkin, YU.M. Kozyrev, A.V. Ponomarev, V.N. Samarov. Osobennosti protsessa uplotneniya poroshkovykh materialov v neodnorodnom nestatsionarnom temperaturnom pole. «Mekhanika kompozitsionnykh materialov i konstruktsiy». [Features of the process of compaction of powder materials in an inhomogeneous nonstationary temperature field. "Mechanics of composite materials and structures."] 2011. v. 17. №3.

[3] Druyanov B.A. Prikladnaya teoriya plastichnosti poristykh tel. [Applied theory of plasticity of porous bodies.] M.: Mashinostroyeniye, 1989.

[4] Grin R.Dzh. Teoriya plastichnosti poristykh tel. [Theory of plasticity of porous bodies] //Sb. perevodov, «Mekhanika», 1973, №4.

Keyword Index

A508 Grade 4N	41
Accident Tolerant Fuel	107
Additive Manufacturing	55,67,78
Alloys	1
Aluminum	101
Bend	101
Bendable	101
Capsule Filling	120,141
Compact Heat Exchangers	17
Composite Component	78
Composite Material	91
Cooling Channel	101
Corrosion Resistance	101
Corrosion	101
Densification	120,160
Density	113
Design Flexibility	101
Design	101
Diffusion Bonding	101
Diffusion Welding	17
Discrete Element Method (DEM)	120,131,141
Distortion	55
DSS	29
Duplex Stainless Steel	29
Electrolyzer	22
F357	55
FeCrAl	107
Finite Element Analysis	141
Finite-Element-Method (FEM)	120
Flexible Tube	101
Flow Rate	101
Flow	101
Fuel Cell	22
GH4169	67
Grain Size	113
Heat Exchange	101
Heat Treatment	48
High Pressure Heat Treatment	55
High Quality Sputtering Target	113
High Strength Aluminum Alloys	55
HIP Modeling	150,166,172
HIP Process	177
HIP Tooling Design	172
HIP-Quench	48
HPHT	55
HSS	91
Hydrogen Blistering	55
Hydrogen	22
Idaho Calcine Simulant	131
Inconel 718	48
Irradiation Embrittlement	85
Large Complex Shape Parts	166
Large HIP	1
Laser Powder Bed Fusion (L-PBF)	48
Low Alloy Steel	41
Magnetically Soft Material	61
Mechanical Properties	11,91
Melting Metallurgy	91
Modeling	1,17,141,160,120
Near-Net-Shape HIP	11,35
Net-Shape	78
Nickel-Based Alloys	48

Non-Uniform Temperature Field 177
NORSOK 35
Nuclear Energy 1
Nuclear Materials 107

PBF-EB 78
Pellet Mills 91
Phase Transformation/precipitation 120
Plasticity 160
PM 29
PM-HIP 41
Porous Structure 22
Powder Bed Fusion – Electron Beam 78
Powder FeSi6.5 61
Powder Filling 131
Powder Hot Isostatic Pressing 22
Powder Material 177
Powder Metallurgy 29,91,107,113
Powder Production 1
Powder Rheology 150
Power Generation 1
Process of Post-Processing 61
Properties 67

Quality Improvement 91

Radioactive Waste 131
Rapid Cooling 48
Recrystallization Embrittlement 85
Rollers 91

Salt 22
sCO2 Turbomachinery 11
SDSS 29
Service Life Extension 91
Shape Holding Insert 22
Simulation 120
Sodium Aluminate 22

Sodium Chloride 22
Stainless Steel (SS) 101
Stainless Steel 316L 160
Super Duplex Stainless Steel 29,35
Superalloy 11

Titanium 22
Tungsten 85
Twin Screw Extruders 91

Uniform Rapid Quenching 55
URQ 55

Viscoplasticity 160

Wear-Resistance 78

Yield Criterion of Green 177

www.ingramcontent.com/pod-product-compliance
Lightning Source LLC
Chambersburg PA
CBHW071214210326
41597CB00016B/1812